SpringerBriefs in Business

More information about this series at http://www.springer.com/series/8860

Marco Fortis · Stefano Corradini
Monica Carminati

Italy's Top Products in World Trade

The Fortis-Corradini Index

Springer

Marco Fortis
Centre of Research in Economic Analysis and International Economic Development (CRANEC)
Università Cattolica del Sacro Cuore
Milan
Italy

Monica Carminati
Fondazione Edison
Milan
Italy

Stefano Corradini
Fondazione Edison
Milan
Italy

ISSN 2191-5482 ISSN 2191-5490 (electronic)
SpringerBriefs in Business
ISBN 978-3-319-15816-7 ISBN 978-3-319-15817-4 (eBook)
DOI 10.1007/978-3-319-15817-4

Library of Congress Control Number: 2015932237

Springer Cham Heidelberg New York Dordrecht London

Printed on acid-free paper

Springer International Publishing AG Switzerland is part of Springer Science+Business Media (www.springer.com)

Preface

This essay analyzes Italy's competitiveness and introduces a new Index compiled by Marco Fortis and Stefano Corradini for Fondazione Edison, which highlights the strong points of Italy's foreign trade, with data updated to 2012 on the trade balance of over 5,000 goods.

Contrary to widespread opinion, Italy is one of the world's most competitive countries with an extraordinary position of leadership in the world trade. This is recognized in the Trade Performance Index (TPI), which was compiled for the first time in 2006 by UNCTAD/WTO's International Trade Centre and is updated every year. This index puts Italy in second place, just behind Germany, in the competitiveness classification of international trade, based on the 14 macro-sectors into which international trade is divided. This is also demonstrated by the Fortis-Corradini Index (FCI) developed in 2010, which provides greater sectorial detail than the TPI by referring to 5,117 products broken down into the six-digit HS 1996 international classification available on the UN Comtrade database. According to this index, for nearly 1,000 products (932 to be exact) Italy in 2012 was either first, second or third worldwide in terms of foreign trade surplus. More specifically, for that year Italy was first for 235 products (with a total positive trade balance of 56 billion dollars for those goods), second for 376 products (for 68 billion dollars) and third for 321 products (for 53 billion dollars) for a total trade surplus of 177 billion dollars. Only three countries (China, Germany and the United States) surpassed Italy in 2012 in terms of the number of first, second and third places in their trade balance worldwide, and only five countries (the three cited above plus Japan and South Korea) registered a higher total value of trade surplus than Italy for the goods in which they place among the top three countries in terms of trade balance (excluding crude oil and natural gas). If, on the other hand, we look at the total number of first, second and third places in the worldwide trade balance of products per 100,000 inhabitants, Italy is second only to Germany. Furthermore, according to the FCI, for about 4,000 internationally traded and statistically surveyed non-food manufactured goods (determined by subtracting energy-related items and agricultural and food products from the 5,117 products that make up the HS96 international classification), Italy has over 2,000 products with a trade surplus and for 1,235 of those, it beats Germany, the benchmark, in terms of positive trade balance. Finally, Italy has 179 mechanical engineering products, out of a total of 496 products in the sector, with a higher trade surplus than Germany's for the same products. In 2012 those 179 products achieved a trade surplus of 32 billion dollars (1.6 % of Italy's GDP). This demonstrates the profound transformation that "Made in Italy" has undergone over the last few years. Indeed, Italy's manufacturing and exports are no longer specialized solely in the traditional sectors of fashion, furniture and food products, but also in mechanical engineering. Other hi-tech sectors like pharmaceuticals are also developing rapidly.

Marco Fortis
Stefano Corradini
Monica Carminati

Contents

Chapter 1
Introduction

Abstract The problem of Italy's low GDP growth is not due to the country's lack of international competitiveness and its inability to compete on global markets, as much as it is a question of weak domestic demand, sapped by twenty years of growing taxation and austerity aimed at reducing Italy's public debt. Indeed, there is extensive recent statistical evidence affirming the high external competitiveness of Italy's industrial manufacturing system: (a) Italy's manufacturing sector is second in Europe and sixth in the world in terms of generated value added; (b) from 1999 to 2013 Italy's share of worldwide exports of manufactured products decreased less than that of other developed countries like the United States, Japan, France and the United Kingdom; (c) in 2012 Italy joined the elite group of economies that boasts a foreign trade surplus of over 100 billion dollars in manufactured goods; (d) in 2013 Italy's trade balance for manufactured goods closed with a surplus of over 98 billion euros, the highest level ever achieved; (e) from 2010 to 2013 Italy improved its total trade balance by 60.5 billion euros, moving from a deficit of 30.1 billion euros to a surplus of 30.4 billion euros; and (f) Italy's 2013 trade surplus of 30.4 billion euros is the fourth largest in the EU or second only to Germany if one excludes the "anomalous" cases of The Netherlands and Ireland (the first being purely a transit country for non-EU goods going to neighboring countries, and the second country a fiscal hub).

Keywords GDP growth · Italy's product specialization · Strong points of Italy's foreign trade · External competitiveness of Italy's manufacturing system

This essay analyzes Italy's competitiveness and introduces a new index compiled by Fondazione Edison which highlights the strong points of Italy's foreign trade, with data updated to 2012 on the trade balance of over 5,000 goods.

In recent years the problem of Italy's and other mature economies' weak GDP growth has often been confusedly associated with low competitiveness and the inability to compete on global markets. In reality over the last 20 years competitiveness in foreign trade has had very little to do, positively or negatively, with the more developed economies' GDP growth, with the exception of the case of Germany. Figure 1.1 clearly illustrates how factors other than competitiveness have been the real

M. Fortis et al., *Italy's Top Products in World Trade*,
SpringerBriefs in Business, DOI 10.1007/978-3-319-15817-4_1

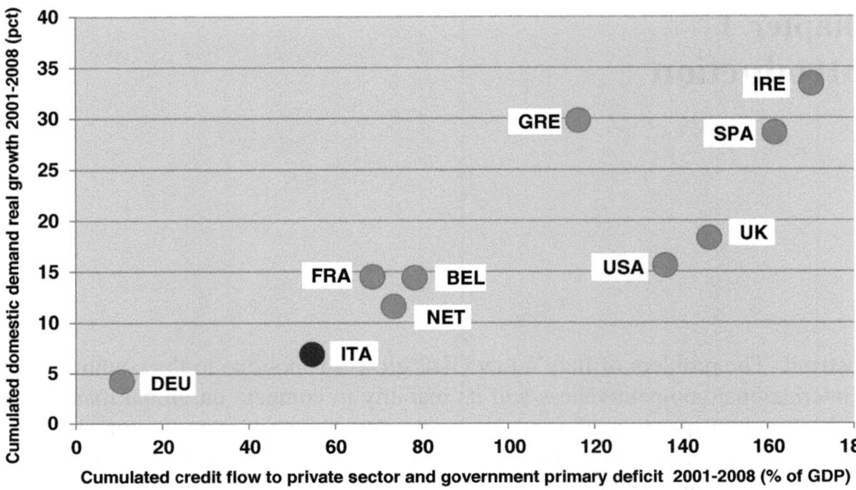

Fig. 1.1 "Drugs", not competitiveness, have been the real drivers of growth in 2001–2008. *Source* compiled by Fondazione Edison on data from Eurostat

drivers of growth in the recent past: in the developed world economies which in the years prior to the crisis had the highest rates of development were driven above all by the unbalanced growth of private debt and not by the performance of their production systems on international markets. Over time, the explosion of private debt fed unsustainable economic growth, based on the well-known real estate and financial "bubble" which exploded at the end of 2008, leading to the most serious economic-financial crisis since that of 1929.

In the years prior to the crisis Italy did not experience this "bubble" in any way. Households' and companies' private debt rose moderately but was still one of the lowest in the world, while the government attempted to stabilize public debt, a heavy legacy of the past, lowering it from 117.2 % of GDP in 1994 to 99.7 % in 2007. Whereas in the pre-crisis years Italy grew only slightly, this did not depend or was only partially a result of a temporary loss of its economic system's competitiveness. Nor was the cause of the country's weak growth an unfavorable product specialization (Tiffin 2014), which would have exposed Italy to emerging market competition, as asserted (erroneously, in our opinion) even by the European Commission (2013, 2014). What actually slowed Italy's growth, in addition to the myriad of country's institutional and infrastructural system constraints that for years have impeded companies and discouraged foreign investment (bureaucracy, fiscal pressure, the high cost of energy, the uncertain legal framework and the infrastructure deficit) was the prolonged austerity that the country was subjected to in an effort to reduce the public debt: while on the one hand the process of deleveraging public finances, based on continuously increasing taxes for households and companies, allowed Italy from 1992 to 2013 to remain in primary surplus for 21 out of 22 years (no other EU country, nor the US or Japan has been able to achieve this!), on the other hand this led to excessive fiscal pressure on the private sector, a reduction in households' disposable incomes with consequent low growth of consumption and a similar negative impact on industrial production and investment. The situation became even more critical in 2011 when excessive austerity policies (with no growth) imposed from Europe dealt a further blow to Italian households' purchasing power. All these factors drastically slowed Italian domestic demand, an important driving of economic growth in any developed economy, especially if the country is a major producer of manufactured goods as is the case of Italy. Thus the absence of an adequate level of domestic consumption, not weak external competitiveness, has been the main cause of Italy's low GDP growth over the last twenty years.

To demonstrate the lack of validity of this widespread belief in Italy's weak external competitiveness we point out how Italian exports reacted to rapidly overcome the temporary steep decline in world trade in 2009: in 2013 Italian exports had already recovered to well above pre-crisis levels, compensating, albeit only partially, for the dramatic drop in domestic demand which weighed negatively on GDP in 2012–14.

After all, if Italy were really as non-competitive as some think, it would not have been able to react, as it actually did, to the fierce asymmetrical competition from Asian countries at the beginning of the new century; to cope with this competition it moved up to segments of higher value-added in the sectors most exposed to Asian competition (personal and household goods) and increased its orientation toward metal products and mechanical engineering (which includes equipment and industrial machinery, household and electrical appliances), means of transport other than automobiles (such as luxury yachts, cruise ships and helicopters), packaged pharmaceuticals, cosmetics and niche chemical products (ICE 2014). In fact, Italy's product specialization today has profoundly changed from what it was 20 years ago (Fortis 1984, 1996, 1998), when fashion and furniture made up three fourths of the foreign trade manufacturing surplus: the weight of the traditional sectors of fashion and furniture in the manufacturing surplus went from 74 % in 1994 to 30 % in 2013 and this 30 %, as mentioned above, is currently composed mainly of higher value added and luxury items compared to the past. The value of Italian exports of mechanical engineering products is currently nearly double that of textiles-clothing-leather-shoes. Furthermore, exports of pharmaceuticals are more than twice that of furniture.

Proof of Italy's dramatic shift towards hi-tech production is the fact that from 2010 to 2013 Italian exports of packaged pharmaceuticals grew by 6.8 billion dollars, the largest increase in the world in absolute value. This result was achieved not only by Italian companies but above all with the significant investment of foreign multinationals, attracted by Italy's production facilities, research centers and a qualified workforce at competitive cost levels: all these factors have made provinces such as Latina, Milan, Frosinone, Bari, Ascoli Piceno and others into absolute leaders in the pharmaceutical industry at an international level, in spite of the many obstacles to doing business that continue to afflict Italy's economic system.

Thus, in spite of austerity policies, chronic inefficiencies on the political, bureaucratic and infrastructural level, asymmetrical Asian competition and the collapse of global trade in 2009, Italy is one of the most competitive countries in the world despite certain indicators put out by organizations like the International Institute for Management Development in Lausanne or the World Economic Forum which describe Italy's chronic decline in the global context (UNIDO 2013; Andreoni 2015). But what the IMD and the WEF are describing are not so much "competitiveness indicators" as "attractiveness indicators" of economic systems and from this perspective Italy does rank low in the international classifications, given the negative factors of public administration, infrastructure, energy and legal uncertainty. The external competitiveness of Italy's industrial-manufacturing system, however, is very high, as shown by numerous recent studies by the Fondazione Edison (Fortis 1998, 2004, 2005, 2006a, b, 2008, 2009, 2011a, b, 2013; Fortis and Quadrio Curzio 2006; Fortis and Carminati 2009, 2012; Fortis and Corradini 2010; Quadrio Curzio and Fortis 2000, 2007, 2012) and as is affirmed by the most recent statistics, which we will briefly summarize below:

- The Italian manufacturing sector in 2011 is second in Europe and sixth in the world in terms of generated value-added (World Bank 2014).
- From 1999 to 2013 Italy's share in world exports of manufactured products decreased, but to a lesser degree than other advanced countries such as the United States, Japan, France and the United Kingdom.
- It should be pointed out that a country's share of world exports is a misleading parameter for measuring countries' international competitiveness: to have a clearer idea one should consider not just exports, but also each country's share of world imports of manufactured products, so in the end the trade balance is the most important indicator of a country's competitiveness. In 2012 Italy joined an elite group of global economies which boasts a trade surplus of manufactured goods (excluding food) of over 100 billion dollars (Fortis 2013). In 2013 Italy's manufacturing surplus (excluding food) reached 131 billion according to the WTO's latest data (WTO 2014). Italy is behind only China (with a trade surplus of 942 billion), Germany (416 billion), Japan (226 billion) and South Korea (218 billion).
- According to ISTAT, in 2013 Italy's manufacturing total trade balance (including also food) closed with a surplus of 97 billion euros, the highest value ever achieved in history; 49 billion euros of this was generated by the non-electronic machinery sector (equipment and industrial machinery), an industry in which Italy boasts the third largest foreign surplus in the world after Germany and Japan.
- Between 2010 and 2013 Italy improved its total trade balance by 60.5 billion euros, going from a deficit of 30.1 billion euros to a surplus of 30.4 billion euros. This is the best result in absolute value in the EU, including Germany, and is 87 % dependent on the growth of exports during this period. Italian imports actually decreased from 2010 to 2013 by just 8 billion euros compared to a 52.5 billion euros increase in exports.
- The Italian total foreign trade surplus of 30.4 billion euros in 2013 is fourth of the EU's or even second only to Germany's if one excludes the "anomalous" cases of

The Netherlands and Ireland (the first being purely a transit country for non-EU goods going to neighboring countries, and the second country a fiscal hub).

Even the Trade Performance Index (TPI) developed by UNCTAD/WTO's International Trade Centre affirms Italy's strength on international markets: this index, which will be briefly described in Chap. 2, places Italy at the top in terms of competitiveness in world trade, just behind Germany. This position is maintained in 2013 (the latest available update) in spite of the drastic slowdown of the Italian economy, which entered recession in 2011 principally because of the collaps of the domestic demand generated by the austerity.

Chapter 3 will introduce a new indicator, the Index of competitive excellence in international trade compiled by Marco Fortis and Stefano Corradini for Fondazione Edison—the Fortis-Corradini Index—which highlights how in recent years, in spite of the global economic crisis, "made in Italy" has achieved extraordinary levels of preeminence on foreign markets. This fact remains unknown to the majority of Italian and international public opinion and this index intends to shed proper light on it.

References

Andreoni A (2015) The industrial competitiveness of Italian manufacturing. A comparison between the UNIDO's Competitive Industrial Performance Index, the WEF and the IMD rankings. Presentation at the International Conference "Where do we stand? Global perspectives on the industrial competitiveness of Italian manufacturing". Milan, 27 January

European Commission (2013) In-depth Review for Italy, 10 April

European Commission (2014) In-depth Review for Italy, 5 March

Fortis M (1984) L'apporto del sistema moda-arredo-casa alla bilancia commerciale italiana. Nomisma, Bologna

Fortis M (1996) Crescita economica e specializzazioni produttive. Sistemi locali e imprese del made in Italy, Milano, Vita e Pensiero

Fortis M (1998) Il made in Italy. Il Mulino, Bologna

Fortis M (2004) La dinamica dei Sistemi locali e dei Distretti italiani: dal territorio ai mercati internazionali. In: Garonna P, Gros-Pietro G M (eds) Il modello italiano di competitività, Ricerca per il Convegno Biennale 2004 del Centro Studi Confindustria, Edizioni Il Sole 24 Ore, Economia e Management

Fortis M (2005) Le due sfide del made in Italy: globalizzazione e innovazione. Profili di analisi della Seconda Conferenza Nazionale sul commercio con l'estero. Fondazione Edison Series. Il Mulino, Bologna

Fortis M (2006a) I distretti produttivi e la loro rilevanza nell'economia italiana: alcuni profili di analisi. In: Fortis M, Quadrio Curzio A (eds) Industria e distretti. Un paradigma di perdurante competitività italiana. Fondazione Edison Series. Il Mulino, Bologna

Fortis M (2006b) L'Italia ai vertici mondiali nelle "4 A". «Quaderno di Approfondimenti Statistici», Fondazione Edison, no. 7, December

Fortis M (2008) Nuovi record per made in Italy e distretti. Nel 2007 il surplus con l'estero delle "4 A" sale a 113 miliardi di euro. «Quaderno di Approfondimenti Statistici», Fondazione Edison, no. 22, April

Fortis M (2009) Mille prodotti vincenti. L'export molecolare fa grande l'Italia. In Il Sole 24 Ore, 29 dicembre

Fortis M (2011a) Esportazioni, investimenti diretti esteri e competitività del sistema Paese nel mercato internazionale. In Economia italiana, 1/2011
Fortis M (2011b) Competitiveness and export performance of Italy. In: Codogno L, Paganetto L (eds) Mesuring Italy's external competitiveness. Rome, Rubettino, pp 367–385
Fortis M (2013) L'Italia nel club dei 100 miliardi per surplus commerciale nei manufatti. «Quaderno di Approfondimenti Statistici», Fondazione Edison, no. 129, August
Fortis M, Quadrio Curzio A (eds) (2006) Industria e distretti. Un paradigma di perdurante competitività italiana. Fondazione Edison Series. Il Mulino, Bologna
Fortis M, Carminati M (2009) Sectors of excellence in the Italian industrial districts. In: Becattini G, Bellandi M, De Propis L (eds) A hand-book of Industrial districts. Edward Elgar, Cheltenham
Fortis M, Carminati M (2012) Il contributo delle «4 A» del made in Italy e dei distretti industriali nella storia dell'export italiano: il caso della meccanica. In: Quadrio Curzio A, Fortis M (eds) L'industria nei 150 anni dell'unità d'Italia. Paradigmi e protagonisti. Fondazione Edison Series. Il Mulino, Bologna
Fortis M, Corradini S (2010) I mille primati del Made in Italy. «Quaderno di Approfondimenti Statistici», Fondazione Edison, no. 47, January
ICE (2014) L'Italia nell'economia internazionale, Rapporto ICE 2013–2014
Quadrio Curzio A, Fortis M (eds) (2000) Il made in Italy oltre il 2000. Fondazione Edison Series. Il Mulino, Bologna
Quadrio Curzio A, Fortis M (eds) (2007) Valorizzare un'economia forte. L'Italia e il ruolo della sussidiarietà. Fondazione Edison Series. Il Mulino, Bologna
Quadrio Curzio A, Fortis M (eds) (2012) L'industria nei 150 anni dell'Unità d'Italia. Paradigmi e protagonisti. Fondazione Edison Series. Il Mulino, Bologna
Tiffin A (2014) European Productivity, Innovation and Competitiveness: the case of Italy. IMF Working Paper, May
UNIDO (2013) The Industrial Competitiveness of Nations. Looking back, forging ahead. Competitive Industrial Performance Report 2012/2013
World Bank (2014) World Development Indicators 2014, Table 4.3
WTO (2014) Statistics Database, http://stat.wto.org

Chapter 2
Italy's Competitiveness According to UNCTAD/WTO's Trade Performance Index

Abstract This chapter briefly describes the Trade Performance Index (TPI) compiled by UNCTAD/WTO's International Trade Centre. The index analyzes the main actors' relative positions in international trade based on a comparison of about 190 countries, with exports broken down into 14 macro-sectors. According to the TPI in 2013 Italy is in second place after Germany in terms of international trade competitiveness, winning 3 first places out of the 14 macro-sectors (in textiles, clothing and leather/footwear), 5 second places (non-electronic machinery, transport equipment, electronic components, basic manufactures and miscellaneous products), and one seventh place (processed foods). Only Germany is ahead of Italy, with 8 first places and one second place.

Keywords Trade Performance Index (TPI) · Italy's external competitiveness · Textiles · Clothing · Leather/footwear · Machinery

The Trade Performance Index (TPI), compiled for the first time in 2006, analyzes the main actors' relative positions in international trade based on a comparison of about 190 countries and the export of goods from 14 macro-sectors in which world trade is divided. These sectors are: Fresh food, Processed food, Wood products, Textiles, Chemicals, Leather products, Basic manufactures, Non-electronic machinery, IT and Consumer electronics, Electronic components, Transport equipment, Clothing, Miscellaneous manufacturing and Minerals. For every macro-sector in each country a composite index was constructed, called the Current Index, based on 5 sub-indicators: (1) net exports; (2) per capita exports; (3) share in world market; (4) product diversification (No. of equivalent products); (5) market diversification (No. of equivalent markets) (ITC 2014). Thus the TPI takes into account not only the absolute value of trade, but also the size of the various countries and their specializations, as well as weaknesses deriving from excessive concentration of exports in a few products or a few target markets.

Table 2.1 summarizes each G-20 country's position in the first 10 places in the world classification of competitiveness of foreign trade in the 14 sectors that make up the TPI. The results of the TPI show Germany's absolute leadership in international trade; in 2013 it won 8 first places and 1 second places for competitiveness

M. Fortis et al., *Italy's Top Products in World Trade*,
SpringerBriefs in Business, DOI 10.1007/978-3-319-15817-4_2

Table 2.1 UNCTAD/WTO Trade Performance Index 2013 (Current index[a]. Ranking of international competitiveness (189 nations); Number of top 10 placings in the world rankings for foreign trade competitiveness in 14 sectors[b])

		Number of best positions	Number of second positions	Number of third positions	Number of fourth positions	Number of fifth positions	Number of sixth positions	Number of seventh positions	Number of eighth positions	Number of ninth positions	Number of tenth positions
1	Germany	8	1								
2	**ITALY**	**3**	**5**					**1**			
3	Russia	1									
4	China		2	1		1	1		2		
5	South Korea			1	1			1	1	1	
6	France			1	1		1				1
7	Turkey			1	1						1
8	Australia				1	1					
9	Japan					1		1	1	1	
10	United States						1		1		
11	India						1				1
12	South Africa						1				
13	Brazil							1			
14	Indonesia								1		
15	Argentina									1	
15	Saudi Arabia									1	
16	United Kingdom										

(continued)

Table 2.1 (continued)

		Number of best positions	Number of second positions	Number of third positions	Number of fourth positions	Number of fifth positions	Number of sixth positions	Number of seventh positions	Number of eighth positions	Number of ninth positions	Number of tenth positions
16	Canada										
16	Mexico										

[a] Sum of 5 sub-indexes: net exports, per capita exports, share in world market, product diversification, market diversification

[b] Fresh food, Processed food, Wood products, Textiles, Chemicals, Leather products, Basic manufactures, Non-electronic machinery, IT and Consumer electronics, Electronic components, Transport equipment, Clothing, Miscellaneous manufacturing, Minerals

Source compiled by Fondazione Edison on data from International Trade Centre UNCTAD/WTO

Table 2.2 Italy's competitiveness according to the Trade Performance Index UNCTAD/WTO, year 2013 (billion dollars)

Sectors	Position of Italy in the world ranking of Trade Performance Index 2013	Value of Italy's export 2013	Italy's trade balance 2013
Clothing	1	23.7	8.0
Leather products	1	24.2	12.3
Textiles	1	13.5	4.9
Non-electronic machinery	2	104.2	70.2
Transport equipment	2	44.5	8.2
Basic manufactures	2	62.0	18.6
Miscellaneous manufacturing	2	49.1	21.3
Electronic components	2	23.2	2.7
Processed food	7	32.0	5.5
Total 9 best sectors		376.4	151.7

Source compiled by Fondazione Edison on data from International Trade Centre UNCTAD/WTO

out of the total of 14 macro-sectors analyzed. Nevertheless, Italy also proved to be very competitive, coming in first 3 times in the UNCTAD/WTO ranking: in textiles, clothing and leather products. What's more, Italy also came in second 5 times: in non-electronic machinery (where it competes almost on equal terms with Germany), transport equipment, electric and electronic components, miscellaneous manufacturing (mainly sunglasses and jewelry) and in basic manufactures (which includes sectors such as metal products, marble and ceramic tiles, where Italy is among the top producers in the world) (Fortis 2008, 2014, 2015). Italy reached the second position in transport equipment and electronic components in 2013. In addition, Italy is seventh place in processed food (which includes wine) (Said 2015). These 9 macro-sectors alone, in which Italy led worldwide in 2013, generated a total of 376.4 billion dollars of exports and a trade surplus of 151.7 billion dollars (Table 2.2).

In particular, among the G-6 countries, China and South Korea, Italy is the country that, along with Germany, holds the highest number of placements worldwide (Table 2.3).

Table 2.3 Position of G-6 Countries, China and South Korea in the ranking of competitiveness of the Trade Performance Index UNCTAD-WTO, year 2013 (ranking in each sector worldwide; in bold the placements among the top 10 most competitive countries)

Countries	Germany	Italy	China	South Korea	Japan	France	United Kingdom	United States
Sectors								
Fresh food	27	37	50	79	89	**6**	42	**8**
Processed food	**1**	**7**	24	71	88	**3**	42	38
Wood and paper	**1**	25	36	50	53	28	35	31
Textiles	**2**	**1**	**3**	**8**	35	19	22	35
Leather products	15	**1**	**2**	38	74	16	21	40
Clothing	15	**1**	**2**	47	79	12	19	41
Chemicals	**1**	28	24	**9**	**8**	**4**	21	**6**
Basic manufactures	**1**	**2**	**6**	**4**	**7**	27	32	47
Non-electronic machinery	**1**	**2**	**5**	11	12	**10**	14	25
Electronic components	**1**	**2**	40	17	**5**	20	26	30
IT and Consumer electronics	12	22	**8**	**7**	40	18	15	23
Transport equipment	**1**	**2**	18	**3**	12	14	34	35
Miscellaneous manufacturing	**1**	**2**	**8**	41	**9**	23	27	25
Minerals	31	46	75	66	85	28	23	21

Source compiled by Fondazione Edison on data from International Trade Centre UNCTAD/WTO

References

Fortis M (2008) L'Italia è seconda per competitività nel commercio mondiale. Il Trade Performance Index UNCTAD/WTO 2006. «Quaderno di Approfondimenti statistici», Fondazione Edison, no. 26, August

Fortis M (2014) Europa e Italia più competitive ma senza domanda interna. «Quaderno di Approfondimenti Statistici», Fondazione Edison, no. 142, November

Fortis M (2015) The Fortis-Corradini Index. Presentation at the International Conference "Where do we stand? Global perspectives on the industrial competitiveness of Italian manufacturing". Milan, 27 January

International Trade Centre—ITC (2014) Trade Competitiveness Map. Benchmarking national and sectorial trade performance. TPI technical notes

Said AJ (2015) The competitiveness of Italian Manufacturing—Outlook and strategic considerations. Presentation at the International Conference "Where do we stand? Global perspectives on the industrial competitiveness of Italian manufacturing". Milan, 27 January

Chapter 3
The Fortis-Corradini Index (FCI)—Fondazione Edison

Abstract The Fortis-Corradini Index (FCI) of competitive excellences in international trade highlights Italy's good performance in international trade. According to this index, in 2012 Italy boasted 932 products (out of a total of 5,117 in which international trade is broken down according to the six-digit HS 1996 international classification) in which it was among the first three countries in the world in terms of foreign trade surplus. More specifically, Italy was first for 235 products (with 56 billion dollars of total trade balance for these goods), second for 376 products (68 billion dollars) and third for 321 products (53 billion dollars) for a total value in terms of foreign trade surplus of 177 billion dollars. Furthermore, according to the FCI, out of 4,000 internationally traded, statistically surveyed non-food manufactured items, Italy has over 2,000 with a trade surplus and for 1,235 of those Italy beats Germany, the internationally recognized benchmark, in terms of trade balance.

Keywords Fortis-Corradini Index (FCI) · International trade · Italy's first places in the world by trade surplus · Italy's second places in the world by trade surplus · Italy's third places in the world by trade surplus · Machine and mechanical appliance sector

3.1 The 2012 Fortis-Corradini Index

Italy's strong position in international trade is also reflected in the Fortis-Corradini Index (FCI) of competitive excellence in international trade. This indicator, with a special algorithm developed by the authors in 2010, measures instantaneously and with a high level of detail the number of products in which each country is first, second or third in terms of its trade balance on a world level (Fortis and Corradini 2010). According to this index, in 2012 Italy had nearly 1,000 products (932 to be exact) in which it placed in the first three positions in terms of foreign trade surplus (Table 3.1).

The FCI is based on information from the UN's database on international trade (UN Comtrade) and other sources such as Eurostat and Istat and is based on 5,117

M. Fortis et al., *Italy's Top Products in World Trade*,
SpringerBriefs in Business, DOI 10.1007/978-3-319-15817-4_3

Table 3.1 Products in which Italy holds the top spot in the world trade balance, year 2012; value in billions of dollars

Italy's positioning in the world trade balance	Number of products	Trade balance of products
First	235	56
Second	376	68
Third	321	53
Total	932	177

Index of Italy's competitive excellence in world trade; Fortis-Corradini's Index © (case study of a total sample of 5,117 products that comprise International Trade)
Source Compiled by Fondazione Edison on data from Istat, Eurostat and UN Comtrade

products from the 6 digit breakdown of the HS 1996 classification. The data presented in this essay are from 2012.

In 2012 Italy was second only to Germany in terms of the total number of first, second and third places in the worldwide trade balance of products for every 100,000 inhabitants, ahead of South Korea and France (Fig. 3.1). In absolute terms, on the other hand, Italy was first place worldwide for 235 products by trade balance (the total value of the balance of these goods: 56 billion dollars), second place in 376 products (68 billion dollars) and third place for 321 products (53 billion dollars). The total: 932 positions of excellence, especially for high value added "niche" products, for a total trade surplus value of 177 billion dollars.

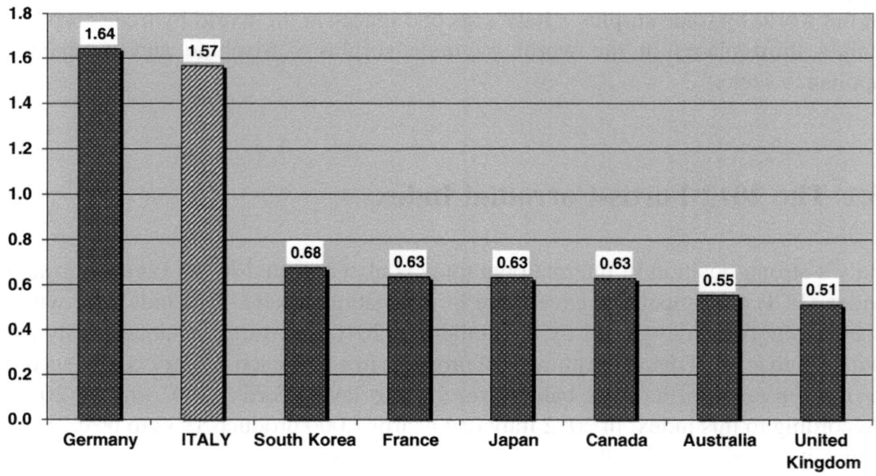

Fig. 3.1 First, second or third place in terms of trade balance, per every 100,000 inhabitants, for each of the G20 countries (Out of a total of 5,117 products). *Source* Processed by Fondazione Edison on data from Istat, Eurostat and UN Comtrade

Only 3 countries (China, Germany and the United States), with population and economy by far larger, performed better than Italy in 2012 in terms of the number of first, second and third places in the trade balance of products worldwide, and only 5 countries (the 3 ones previously mentioned plus Japan and South Korea) registered total values of trade surpluses higher than Italy's for the goods in which they are among the first three countries by trade surplus (excluding crude oil and natural gas). "Made in Italy's" excellence is rounded out with 500 other products in which Italy in 2012 came in fourth or fifth in terms of its worldwide trade balance; these products added another 40 billion dollars to Italy's trade balance surplus.

The strength of "made in Italy" derives from the extensive diversification of its specializations, driven mainly by the "4 F" macro-sectors (Food and wine; Fashion and luxury; Furniture and building materials; Ferrari cars and the whole mechanical engineering industry), but also by other important sectors such as metallurgy, paper and chemicals-pharmaceuticals.

Thousands of medium-large, medium and small enterprises are producing these results and enabling Italy to compete with countries that have many more large-scale, multinational groups, but which do not have Italian companies' typical ability to be flexible and handle hundreds of different types of products, a sort of customizing excellence. It is in this area that the winning factors of "made in Italy" emerge, such as creativity, innovation, quality, design and a strong tradition of "industrial craftsmanship", in other words the ability to build customized products for the clients, even in hi-tech sectors such as mechanical engineering or transportation vehicles.

Italy, like other countries, is suffering from the worst global crisis since 1929. But it would be difficult for this crisis to destroy Italy's ability to practice those traditional and new "trades" that it does better than other countries and at which it excels internationally.

3.2 Products in Which Italy Holds First Place in the World by Trade Surplus

According to the FCI there are 235 products for which Italy in 2012 held first place in the world by trade surplus. Italy's trade balance for these 235 products was 55.7 billion dollars, divided up as follows (Fig. 3.2): 25.6 billion dollars in trade balance was generated by goods in the mechanical engineering-rubber and plastic sectors; 18.4 billion dollars of goods in the sector of fashion and luxury; 7.3 billion dollars of goods in the food and wine sector; 0.4 billion dollars of furniture and building materials and finally 4 billion dollars of other products including goods in the paper industry, glass and chemicals.

There are no sectors in which Italy did not hold a significant number of first places in its trade balance in 2012. For space reasons in Table 3.2 we chose only a few of the most significant products from each sector. Let us examine them briefly.

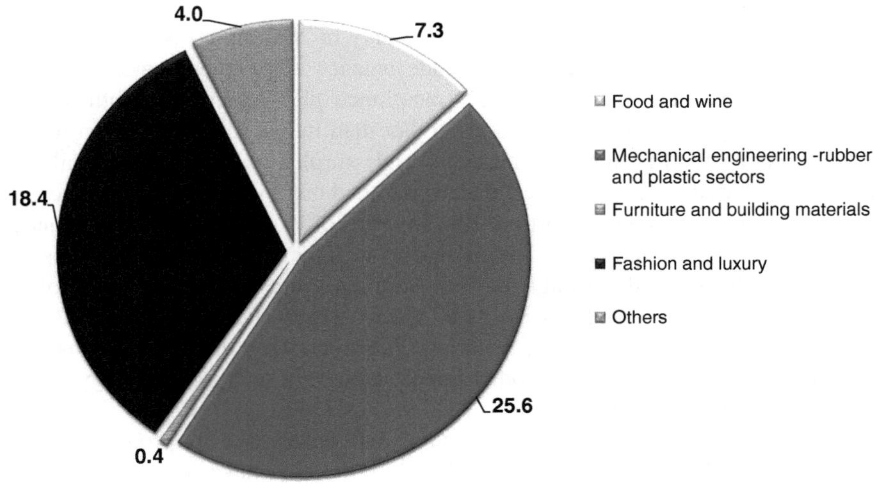

Fig. 3.2 Distribution of 55.7 billion dollars in trade balance generated by the 235 products for which Italy is a world leader: year 2012 (billions of dollars). *Source* Processed by Fondazione Edison on data from Istat, Eurostat, UN Comtrade

Italy leads in a wide range of products from machines and technology for agriculture and tobacco to food products and beverages (Italy leads worldwide in the export of pasta, chocolate and other processed foods containing cocoa, tomato derivatives, apples and salami and seasoned meats) to many types of food processing machines. Other Italian winners are fashion (with many textile-clothing products, leather goods, shoes, jewelry, sunglasses, components for these sectors and industrial machines to manufacture them).

Furthermore, Italy has many winners by global trade balance in the paper products industry (from tissues to paper towels to paper bags), as well as metallurgical products (from iron and steel pipes to aluminum castings) and it also leads in many important metallurgy technologies, heating, cooling, refrigeration and furnace technologies (including commercial refrigeration equipment), as well as machines for woodworking and processing of non-metallic minerals (such as ornamental stones and ceramics). Italy is also first by trade balance for different types of metal products, special hi-tech mechanical machinery (including packaging machinery, metal-working and plastic material machinery). It also holds first place by trade balance for transport equipment, precision and safety products (such as helicopters, insulated copper wires and safety glass for vehicles and parts for security and control equipment), as well as sports and entertainment goods (such as hunting rifles, ski boots and snow-surfing footwear).

Italy also has numerous internationally top positions by trade surplus in rubber and plastic articles such as could not fail in the country that invented polypropylene by Italian Nobel prize winner Giulio Natta. Among these, the records in trade surplus of polypropylene plates, sheets, film and tape production. Some of Italy's more curious

Table 3.2 The products in which Italy holds the top spots in its trade balance with foreign countries: selection of some significant cases for each product category: year 2012 (trade balance in millions of dollars)

Categories and products	Trade balance
Technologies of agriculture and tobacco	
Machinery for preparing or making up tobacco	314
Parts for soil preparation or cultivation machinery	261
Rollers, soil preparation, cultivation machinery, nes	172
Dairy machinery	57
Food products	
Uncooked pasta, not stuffed or prepared, without eggs	1,853
Chocolate/cocoa food preparations nes	1,007
Tomatoes, whole/pieces, prepared/preserved, no vinegar	998
Apples, fresh	910
Swine meat, salted/dried/smoked not ham/shoulder/belly	632
Stuffed pasta	377
Beverages and condiments	
Vinegar and substitutes for vinegar from acetic acid	263
Vermouth and other flavoured grape wines—pack <2 l	204
Technologies of food, beverages and tobacco	
Bakery and pasta making machinery	708
Automatic beverage-vending machines: incorporating heating or refrigerating devices	181
Machinery for preparation of fruits, nuts, vegetables	120
Bakery ovens, etc., non-electric	114
Presses, crushers etc., for wine, fruit juice, beverages	79
Textiles and clothing	
Woven fabric, >85 % combed wool or fine hair, <300 g/m^2	581
Womens, girls ensembles, material nes, not knit	517
Womens, girls dresses, of material nes, not knit	367
Woven fabric, >85 % combed wool or fine hair, >300 g/m^2	146
Woven fabric of silk, nes	72
Fashion accessories	
Shoes and boots, with outer soles and uppers of leather	3,271
Leather and tanned skins of each type	2,671
Handbags with outer surface of leather	2,536
Sunglasses	1,828
Belts and bandoliers of leather or composition leather	367
Frames and mountings for spectacles etc., of plastic	329
Watch straps etc., and parts, of/clad with precious metal	234
Watch cases of, or clad with, precious metal	123

(continued)

Table 3.2 (continued)

Categories and products	Trade balance
Technologies and components of fashion	
Parts of footwear nes, gaiters and leggings etc.	327
Clasps/buckles, etc., for clothing, footwear, bags etc.	271
Machines and parts of machines for the industry of leather and footwear	213
Circular knitting machines, diameter <165 mm	150
Synthetic organic tanning substances. Finishing agents and dye carriers for textile industry	147
Products for home	
Wallpaper and similar wall coverings, nes	74
Materials and mechanical products for the construction industry	
Door closures, automatic, of base metal	228
Valves, pressure reducing	225
Worked calcareous stone nes	77
Paper and its technologies	
Machinery for making up pulp, paper, paperboard nes	495
Paper handkerchiefs, cleansing, facial tissues, towels	422
Sacks and bags, of paper, having a width >40 cm	148
Metallurgy	
3 different types of pipes and hollow profiles of iron, steel and special steels	3,139
11 different types of rods, profiles and wire, of iron and non-alloy steel and semi-finished products, flat products, bars and rods of stainless steel	1,780
Technologies of metallurgy and metalworking	
Parts of metal rolling mills and rolls	634
Boring-milling machines num controlled for metal	264
Calendering or rolling machines, not for metals/glass	167
Tube mills, metal rolling	90
Way-type unit head machines, metal working	28
Metal Products	
Articles, iron or steel nes, forged/stamped, nfw	722
Expanded metal, i/nas <3 mm wire, <100 cm mesh	44
Copper screws, bolts or nuts except wood screws	38
Heating, cooling, refrigeration and furnace technologies and products	
Refrigerator/freezer chests/cabinets/showcases	852
Commercial equipment, hot drinks/cooking/heating food	747
Industrial furnace, oven, incinerator non-electric nes	166
Furnace burners for liquid fuel	92
Special machines for wood and non-metallic mineral products	
Grinding or polishing machines for working stone, ceramics, concrete or similar mineral materials or for cold working glass	794
Machine tools for wood, cork or hard plastic, etc., nes	469

(continued)

Table 3.2 (continued)

Categories and products	Trade balance
Drilling or morticing machines for wood, etc.	190
Grinding, sanding, polishing machines for wood, etc.	58
Mechanics hi-tech	
Packing or wrapping machinery nes	2,403
Pumps nes	969
2 different types of machines for the processing of metals to deformation; machines for bending, folding, straightening, flattening not numerically controlled, shears different from those not combined punching and shearing machines, not numerically controlled	310
Rubber or plastic moulding and forming machines nes	261
Machines for manufacturing or hot working glass or glassware: other	145
Transport equipment and other advanced technologies	
Helicopters of an unladen weight >2,000 kg	1,362
Insulated winding wire of copper	570
Safety glass (tempered) for vehicles, aircraft, etc.	218
Machines for balancing mechanical parts, nes	125
Electric signal, safety and traffic controller parts	83
Goods for fun and sport	
Shotguns, shotgun-rifles for sport, hunting or target	231
Ski-boots, cross-country, ski footwear and snowboard boots	116
Articles of plastics and rubber	
Sheet/film not cellular/reinf polymers of propylene	627
Self-adhesive plastic, rolls <20 cm wide	561
Rubber tube, pipe, hose, metal reinforced, no fittings	208
Camel-back strips for retreading rubber tyres	45
Hard rubber (e.g. ebonite) in all forms, articles, scrap	40
Various food and industrial products	
Play, fair-ground equipment, travelling circus, theatre	172
Beans, shelled, prepared/preserved, not frozen/vinegar	169
Chicory, fresh or chilled, except witloof	58
Buttons, nes	46
Other products	
Domestic iron/steel solid fuel appliances, not cookers	264
Asphalt or similar material articles, in rolls	252
Corks, crown, of base metal	124
Built-in jacking systems for garages	119
Automatic vending machines	83

Source Compiled by Fondazione Edison on data from Istat, Eurostat and UN Comtrade

leading products by trade balance are merry-go-rounds, beans, buttons and chicory, heaters and fireplaces, asphalt works, lifts for auto repair shops and crown corks.

The individual products in which Italy leads internationally in terms of trade balance value are presented in Table 3.3. The category footwear with natural leather soles takes first place overall with 2.6 billion dollars of trade balance in 2012. This is followed in value terms in the trade balance internationally by leather handbags (2.5 billion), packing or wrapping machinery (2.4 billion), pasta (1.85 billion), sunglasses (1.83 billion) and bovine or equine leather (1.7 billion). It is notable that most of these leading products, like the case of footwear, are produced in regional industrial clusters, that are a typical feature of the Italian economy (Fortis and Quadrio Curzio 2006).

For many of the products in which Italy holds leading positions by trade balance (Table 3.9 statistical appendix) Italy positions itself at the highest level of value-added, a level where emerging market manufacturers like China struggle to reach and where the distinctive features of made in Italy in terms of quality, design, innovation and customer service continue to make the difference. For footwear and leather boots, for example, Italy is the uncontested leader in global exports ahead of Spain and Portugal, since Chinese producers face technical difficulties (leather sole processing)

Table 3.3 The top 20 products in which Italy holds the top spot in the world trade balance: year 2012

Rank	Commodity description	Trade balance value (billion $)
1	Footwear, outer soles and uppers of leather, nes	2.65
2	Handbags with outer surface of leather	2.54
3	Packing or wrapping machinery nes	2.40
4	Uncooked pasta, not stuffed or prepared, without eggs	1.85
5	Sunglasses	1.83
6	Bovine and equine leather, full or split grain, nes	1.75
7	Parts of wash, filling, closing, aerating machinery	1.39
8	Helicopters of an unladen weight >2,000 kg	1.36
9	Pipes and tubing, stainless steel, welded	1.11
10	Hollow profiles/tubes, iron/steel, non-circular, welded	1.08
11	Chocolate/cocoa food preparations nes	1.01
12	Tomatoes, whole/pieces, prepared/preserved, no vinegar	1.00
13	Pumps nes	0.97
14	Pipes etc., nes, iron/steel welded nes, diameter <406.4 mm	0.95
15	Bovine and equine leather, nes	0.92
16	Apples, fresh	0.91
17	Refrigerator/freezer chests/cabinets/showcases	0.85
18	Commercial equipment, hot drinks/cooking/heating food	0.75
19	Articles, iron or steel nes, forged/stamped, nfw	0.72
20	Bakery and pasta making machinery	0.71

Index of Italy's competitive excellence in world export markets; Fortis-Corradini's Index (case study of a total sample of 5,117 products that comprises International Trade)
Source Compiled by Fondazione Edison on data from Istat, Eurostat and UN Comtrade

and price obstacles (the cost of the leather) in producing this type of footwear. The same goes for other fashion products that require special attention as far as choice of raw materials, design and manufacturing such as the combination of textile materials for men and women and women's wool suits. Also for non-electronic mechanical engineering, from industrial machinery to packing machinery, Chinese competition is undoubtedly growing in the lower value-added segments but it then runs into difficulty with higher quality, more technologically complex products, where Italy remains the world leader.

3.3 Products in Which Italy Holds Second Place in the World by Trade Surplus

We should not forget Italy's second place products by trade surplus worldwide (Table 3.4), which are often actually in positions of authentic co-leadership rather than subordinate ones. Italy's second most important product category in terms of

Table 3.4 The top 20 products in which Italy holds the second position in the world trade balance: year 2012

Rank	Commodity description	Trade balance value (billion $)
1	Taps, cocks, valves and similar appliances, nes	4.93
2	Grape wines nes, fortified wine or must, pack <2 l	4.56
3	Furniture, wooden, nes	1.88
4	Furniture parts nes	1.87
5	Articles of iron or steel, nes	1.86
6	Wheeled tractors nes	1.56
7	Unglazed ceramic flags, tiles >7 cm wide	1.53
8	Parts of gas turbine engines except turbo-jet/prop	1.44
9	Machinery to fill, close, aerate, etc., bottle, container	1.43
10	Electric conductors, 80–1,000 V, no connectors	1.30
11	Articles of aluminium, nes	1.21
12	Parts of machines and mechanical appliances nes	1.16
13	Motorboats, other than outboard motorboats	1.12
14	Coffee, roasted, not decaffeinated	1.06
15	Parts of vacuum pumps, compressors, fans, blowers, hoods	0.82
16	Communion wafers, rice paper, bakers wares nes	0.76
17	Grapes, fresh	0.72
18	Heat exchange units, non-domestic, non-electric	0.68
19	Machinery for treatment by temperature change nes	0.67
20	Grape wines, sparkling	0.64

Index of Italy's competitive excellence in world export markets; Fortis-Corradini's Index (case study of a total sample of 5,117 products that comprises the International Trade)
Source Compiled by Fondazione Edison on data from Istat, Eurostat and UN Comtrade

trade balance is taps, cocks and valves (4.9 billion dollars), a sector whose strength lies in the dynamism of numerous medium- and medium-large companies in two important regional clusters: Lago d'Orta-Valsesia in Piedmont and Lumezzane-Brescia in the province of Brescia. This is followed by the export of the wine and sparkling wine industry (4.5 billion dollars) in which Italy is preceded only by France. Next are two furniture categories, wooden furniture and furniture parts, followed in the ranking by iron and steel works, agricultural tractors, ceramic flags and tiles, parts of gas turbine engines, machinery to fill, close, label and aerate bottles and containers, aluminum parts, parts of machines and mechanical appliances, motorboats and yachts.

Other important second place products for Italy's international trade balance which are not found in Table 3.4 (Table 3.10 statistical appendix) include extra virgin olive oil, light fixtures, chairs and easy chairs, and polished granite. Other significant products to note are various plastic goods (from ethylene pipes to plastic bathtubs and showers), numerous types of fabrics and garments, metallurgical products, equipment and machinery.

3.4 Products in Which Italy Holds Third Place in the World by Trade Surplus

Products coming in third place in the trade balance must also be mentioned, as they demonstrate Italy's high degree of international specialization (Table 3.5). Italy is third worldwide in the trade balance of jewelry, parts for tractors and personal transport vehicles, flooring tiles and sheets, mechanical machinery and appliances, plastic material products, gearing, ball screws, speed changers and torque converters, air pumps and compressors, couches and easy chairs, brake system parts except linings for motor vehicles, drive axles with differential for motor vehicles, and iron and steel structures and parts of structures. Italy is also technically third place in footwear, sole rubber and plastic uppers of leather behind China and Hong Kong, but in this case it is actually second because most of Hong Kong's exports are re-exports of products from China. Italy is also third in non-office metal furniture, in handbags with exterior in plastic or fabric, and glasses frames (Table 3.11 statistical appendix).

3.5 "Collecting Medals" in International Trade

International trade is a bit like a large multidisciplinary sports event with various countries competing to win as many medals as possible. Like in the Olympics, with the trade surplus we can set up a sort of "medals collection" that keeps track of the number of gold, silver and bronze medals that each country has won in its world trade balance.

Table 3.5 The top 20 products in which Italy holds the third position in the world trade balance: year 2012

Rank	Commodity description	Trade balance value (million $)
1	Jewellery and parts of precious metal except silver	4.28
2	Motor vehicle parts nes	3.79
3	Glazed ceramic flags, tiles wider than 7 cm	2.31
4	Machines and mechanical appliances nes	2.20
5	Plastic articles nes	1.42
6	Gearing, ball screws, speed changers, torque converter	1.37
7	Air or gas compressors, hoods	1.26
8	Seats with wooden frames, upholstered nes	1.22
9	Brake system parts except linings for motor vehicles	1.15
10	Drive axles with differential for motor vehicles	1.01
11	Structures and parts of structures, iron or steel, nes	0.94
12	Footwear, sole rubber, plastics uppers of leather, nes	0.92
13	Bar/rod, i/nas, indented or twisted, nes	0.86
14	Bolts/screws nes, with/without nut/washer, iron/steel	0.84
15	Other flat, not in coils, not further worked than hot-rolled <10 mm	0.78
16	Kitchen furniture, wooden, nes	0.74
17	Automatic washing machines, of a dry capacity <10 kg	0.70
18	Plastic sheet, film, foil or strip, nes	0.69
19	Special purpose motor vehicles nes	0.68
20	Chandeliers, other electric ceiling or wall lights	0.64

Index of Italy’s competitive excellence in the world export market; Fortis-Corradini’s Index (case study of a total sample of 5,117 products that comprises the International Trade)
Source Compiled by Fondazione Edison on data from Istat, Eurostat and UN Comtrade

Based on the FCI, for a total of 5,117 products, Italy in 2012 was the fifth country for the number of first places by world trade balance, the third country for number of second places and the third country for number of third places as well. Considering the total number of first, second and third places in the world trade balance held by all G20 countries, Italy ranks fourth overall (with 932 winning places), after China, Germany and the United States (Table 3.6).

Italy placed sixth in terms of the total value of first, second and third places among the G20 countries, excluding oil, gas and coal. It is preceded by China, Germany, Japan, the United States and South Korea (Table 3.7).

Table 3.6 G20 countries: "Collecting medals" in international trade: year 2012

Rank	Countries	Number of first positions	Number of second positions	Number of third positions	Total number of first, second and third positions
1	China	1,423	468	298	2,189
2	Germany	525	461	335	1,321
3	United States	405	353	309	1,067
4	**ITALY**	**235**	**376**	**321**	**932**
5	Japan	303	280	223	806
6	India	139	209	166	514
7	France	88	153	176	417
8	South Korea	77	140	121	338
9	United Kingdom	65	106	155	326
10	Indonesia	52	95	102	249
11	Turkey	54	94	96	244
12	Canada	78	60	81	219
13	Mexico	49	84	86	219
14	Russia	54	60	62	176
15	Brazil	39	50	54	143
16	Australia	37	54	35	126
17	South Africa	32	37	27	96
18	Saudi Arabia	26	18	22	66
19	Argentina	15	22	23	60
Total		3,696	3,120	2,692	9,508

Number of products in which each G20 country holds first, second and third place in the foreign trade surplus: (out of a total number of 5,117 products)
Positioning of Italy is emphasized in *bold*
Source Compiled by Fondazione Edison on data from Istat, Eurostat and UN Comtrade

3.6 Italy Beats Germany in Over 1,200 Manufactured Products in Terms of Its International Trade Surplus

There are two common clichés about Italian industry: the first is that it is non-competitive and the second that its companies are too small to compete on the new global markets. The main indicator that provides the basis for these two presumed "truths" is that in recent years Italy's market share of worldwide exports has decreased slightly (as has also been experienced by France, Great Britain, the United States and Japan due to the aggressive rise of China and South Korea).

Table 3.7 G20 countries: "Collecting medals" in international trade: year 2012 (billion dollars)

	Countries	Trade balance value of the products in which the G20 countries are first in the world ranking	Trade balance of the products in which the G20 countries are second in the world ranking	Trade balance of the products in which the G20 countries are third in the world ranking	Trade balance of the products in which the G20 countries are first, second and third in the world ranking (excluding crude oil, natural gas and coal)
1	China	1,256	119	34	1,410
2	Germany	250	218	58	526
3	Japan	312	119	53	484
4	United States[a]	240	86	41[a]	353
5	South Korea	105	82	98	285
6	**ITALY**	**56**	**68**	**53**	**177**
7	Russia[b]	206[b]	203[c]	10	176
8	India	51	91	15	158
9	Australia[c]	133[d]	17	30	137
10	Brazil	55	68	10	134
11	France	72	28	20	120
12	Mexico	63	20	22	105
13	Canada	51	19	33	102
14	Indonesia[d]	48	36	19[e]	90
15	United Kingdon	26	16	20	62
16	Turkey	17	13	11	41
17	South Africa	15	21	4	39
18	Saudi Arabia[e]	334[f]	6	3	38
19	Argentina	17	3	14	34
Total		3,308	1,233	549	4,471

Value of the foreign trade balance of the products in which each G20 country are first, second or third in the world ranking of the trade balance for a total of 5,117 products

[a] 14 billion dollars of coal

[b] 62 billion dollars of natural gas

[c] 181 billion dollars of crude oil

[d] 43 billion dollars of coal

[e] 13 billion dollars of coal

[f] 305 billion dollars of crude oil

Positioning of Italy is emphasized in *bold*

Source Compiled by Fondazione Edison on data from Istat, Eurostat and UN Comtrade

But everything changes if one considers trade relations minus energy and primary agricultural commodities and takes as a yardstick of competitiveness not exports but the trade balance of non-food manufactured products as calculated by the World Trade Organization. Indeed, Italy is one of just 6 countries in the G-20 (including China, Germany, Japan, South Korea and India, the latter only from 2013) to have a structural trade surplus in manufactured products of 131 billion dollars in 2013.

But there is another figure that shows Italy's strength on international markets. According to the FCI, out of about 4,000 non-food manufactured products traded internationally and statistically surveyed (determined by subtracting items comprising energy and agricultural and food products from the 5,117 products which make up the international HS96 classification), Italy boasts over 2,000 products with a surplus in the trade balance and for 1,235 of those, it's surplus is bigger than Germany's as a benchmark. These 1,235 products in which Italy "beats" Germany in 2012 (Table 3.12 statistical appendix) achieved a trade surplus of 146 billion dollars (7.2 % of Italy's GDP). These products include gold and jewelry items, taps and related products, tractor and auto vehicle parts and accessories, footwear with outer soles and uppers of leather, leather handbags, packing and wrapping machines, ceramic tiles, iron and steel furniture and works, and sunglasses (Table 3.8). Only China has a greater number of cases in which it is more competitive than Germany: a total of 2,169 products (with a surplus of 16.4 % of China's GDP). The United States is in third place (with 1,071 products but with a surplus of only 1.6 % of its GDP) whereas Japan is in 4th place (Japan beats Germany in a total of 1,065 products for a value of 7.6 % of Japan's GDP) (Fig. 3.3) (Fortis 2013).

3.7 Italy Beats Germany in About 200 Products in the Machine and Mechanical Appliance Sector

As we have seen, over the last decade Italy's manufacturing has increasingly focused on mechanical engineering. Italy's mechanical engineering, contrary to the adage that Italy excels only in more traditional low-tech sectors, is experiencing rapid and significant growth: in 2013 exports of this sector alone made up 13 % of Italy's total exports, affirming the increasing importance of specializing in so-called "medium tech" products which in reality are much more "hi-tech" than what is considered as such in the most commonly used classifications.

According to the FCI there are 496 non-food manufactured products belonging to the machinery and mechanical appliance industry. And out of the over 2,000 products for which Italy has a trade surplus, 405 belong to this industry for a total value of 66 billion dollars. Whereas Germany is considered the most competitive country in the world for foreign trade, it is interesting to note that Italy has 179 "mechanical" products with a higher trade surplus than that of Germany for the same products (Fig. 3.4). In 2012 these 179 products achieved a trade surplus of 32 billion dollars (1.6 % of Italy's GDP).

Table 3.8 Trade balance at world level: the top 30 products in which Italy beats Germany: year 2012 (million dollars)

Rank	Product description	Italy trade balance	Germany trade balance
1	Gold in unwrought forms non-monetary	5,066.9	4,095.5
2	Taps, cocks, valves and similar appliances, nes	4,932.7	3,279.3
3	Jewellery and parts of precious metal except silver	4,281.9	−156.9
4	Motor vehicle parts nes	3,786.9	2,092.2
5	Footwear, outer soles and uppers of leather, nes	2,652.2	−169.8
6	Handbags with outer surface of leather	2,536.0	−227.1
7	Packing or wrapping machinery nes	2,403.2	1,900.7
8	Glazed ceramic flags, tiles wider than 7 cm	2,311.0	−300.1
9	Furniture, wooden, nes	1,877.9	−690.1
10	Furniture parts nes	1,870.7	169.9
11	Articles of iron or steel, nes	1,857.0	1,407.6
12	Sunglasses	1,828.4	−98.7
13	Bovine and equine leather, full or split grain, nes	1,749.0	−37.8
14	Unglazed ceramic flags, tiles >7 cm wide	1,531.8	1.6
15	Parts of gas turbine engines except turbo-jet/prop	1,438.8	1,048.6
16	Parts of wash, filling, closing, aerating machinery	1,393.2	982.6
17	Helicopters of an unladen weight >2,000 kg	1,361.6	198.4
18	Electric conductors, 80–1,000 V, no connectors	1,297.6	186.5
19	Seats with wooden frames, upholstered nes	1,217.3	−1,022.0
20	Articles of aluminium, nes	1,211.2	30.1
21	Parts of machines and mechanical appliances nes	1,164.5	785.2
22	Brake system parts except linings for motor vehicles	1,147.0	913.0
23	Pipes and tubing, stainless steel, welded	1,108.7	25.7
24	Hollow profiles/tubes, iron/steel, non-circular, welded	1,081.0	−447.8
25	Flat otherwise plated or coated with zinc: width > 600 mm	1,053.9	−549.3
26	Pumps nes	968.5	−34.4
27	Pipes etc., nes, iron/steel welded nes, diameter <406.4 mm	949.0	−122.2
28	Structures and parts of structures, iron or steel, nes	938.7	753.1
29	Bovine and equine leather, nes	921.6	−10.3
30	Footwear, sole rubber, plastics uppers of leather, nes	918.4	−940.8

Fortis-Corradini Index, © Fondazione Edison. Analysis on the basis of about 4,000 manufactured products excluding food and energy
Source Compiled by Fondazione Edison on data from Istat, Eurostat and UN Comtrade

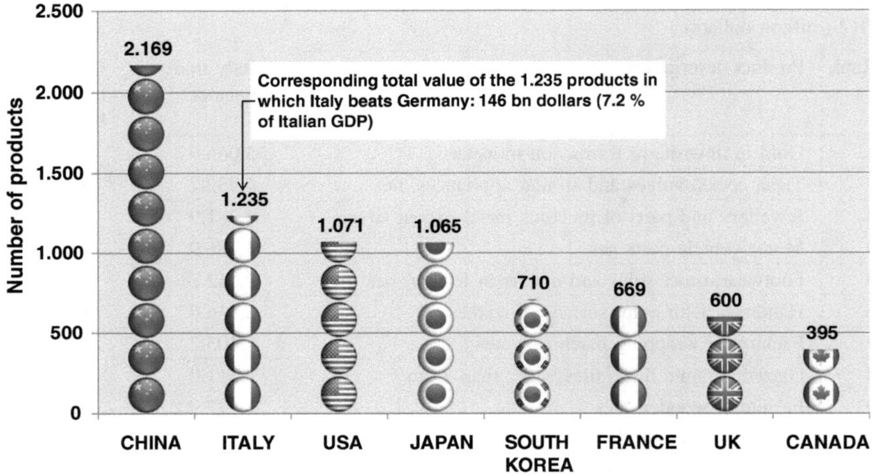

Fig. 3.3 Trade Balance, 2012. Number of products in which each country beats Germany. (Fortis-Corradini Index. Analysis on the basis of about 4,000 products for each country, excluding food and energy) *Source* Compiled by Fondazione Edison on data from UN Comtrade

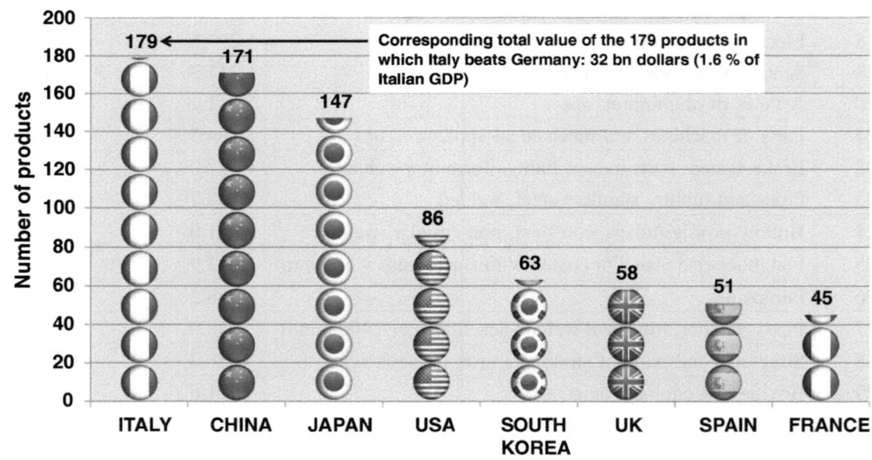

Fig. 3.4 Trade Balance, 2012. Number of products in which each country beats Germany in Machinery and Mechanical Appliances Sector (Fortis-Corradini Index. Analysis on the basis of about 496 products for each country, excluding computer sector) *Source* Compiled by Fondazione Edison on data from UN Comtrade

3.8 Appendix

See Tables 3.9, 3.10, 3.11 and 3.12.

Table 3.9 Products in which Italy holds first place in the world trade balance: year 2012

Rank	Commodity description	Trade balance value (million $)
1	Footwear, outer soles and uppers of leather, nes	2,652.2
2	Handbags with outer surface of leather	2,536.0
3	Packing or wrapping machinery nes	2,403.2
4	Uncooked pasta, not stuffed or prepared, without eggs	1,853.1
5	Sunglasses	1,828.4
6	Bovine and equine leather, full or split grain, nes	1,749.0
7	Parts of wash, filling, closing, aerating machinery	1,393.2
8	Helicopters of an unladen weight >2,000 kg	1,361.6
9	Pipes and tubing, stainless steel, welded	1,108.7
10	Hollow profiles/tubes, iron/steel, non-circular, welded	1,081.0
11	Chocolate/cocoa food preparations nes	1,007.0
12	Tomatoes, whole/pieces, prepared/preserved, no vinegar	998.2
13	Pumps nes	968.5
14	Pipes etc., nes, iron/steel welded nes, diameter <406.4 mm	949.0
15	Bovine and equine leather, nes	921.6
16	Apples, fresh	909.7
17	Refrigerator/freezer chests/cabinets/showcases	852.0
18	Commercial equipment, hot drinks/cooking/heating food	747.1
19	Articles, iron or steel nes, forged/stamped, nfw	721.6
20	Bakery and pasta making machinery	707.7
21	Stainless steel bar nfw than cold formed/cold finished	653.0
22	Parts of metal rolling mills and rolls	633.8
23	Swine meat, salted/dried/smoked not ham/shoulder/belly	631.8
24	Sheet/film not cellular/reinf polymers of propylene	626.8
25	Footwear, soles, uppers of leather, over ankle, nes	618.3
26	Woven fabric, >85 % combed wool or fine hair, <300 g/m^2	580.6
27	Insulated winding wire of copper	569.5
28	Self-adhesive plastic, rolls <20 cm wide	560.5
29	Womens, girls ensembles, material nes, not knit	516.6
30	Machines to agglomerate, shape, mould minerals or fuel	506.2
31	Machinery for making up pulp, paper, paperboard nes	494.7
32	Machine tools for wood, cork or hard plastic, etc., nes	468.7
33	Paper handkerchiefs, cleansing, facial tissues, towels	422.2
34	Stuffed pasta	376.5
35	Belts and bandoliers of leather or composition leather	367.4
36	Womens, girls dresses, of material nes, not knit	366.6

(continued)

Table 3.9 (continued)

Rank	Commodity description	Trade balance value (million $)
37	Machine tools nes for stone, ceramics and cold glass	353.7
38	Parts, accessories nes, metal shaping machine tools	350.9
39	Cranes designed for mounting on road vehicles	332.3
40	Frames and mountings for spectacles etc., of plastic	328.7
41	Parts of footwear nes, gaiters and leggings etc.	326.8
42	Polyesters nes, unsaturated, in primary forms	315.3
43	Machinery for preparing or making up tobacco	313.8
44	Textured yarn nes, nylon, polyamide >50 dtex not retail	303.7
45	Woven fabric, carded wool or hair + manmade staple fib	288.3
46	Patent leather, metallised leather	285.9
47	Womens, girls overcoats, of impregnated, etc., fabric	280.2
48	Clasps/buckles, etc., for clothing, footwear, bags, etc.	270.6
49	Garments of knit or crochet impregnated fabric	265.8
50	Boring-milling machines num controlled for metal	264.2
51	Domestic iron/steel solid fuel appliances, not cookers	263.6
52	Vinegar and substitutes for vinegar from acetic acid	263.4
53	Rubber or plastic moulding and forming machines nes	260.8
54	Parts for soil preparation or cultivation machinery	260.6
55	Nonwovens, man-made fila	258.0
56	Asphalt or similar material articles, in rolls	252.1
57	Cheese, grated or powdered, of all kinds	249.1
58	Bars and rods, iron, n/f/workd	246.4
59	Grinding/polishing machines for stone, ceramics, glass	242.2
60	Dish washing machines commercial	235.7
61	Watch straps etc., and parts, of/clad with precious metal	233.9
62	Shotguns, shotgun-rifles for sport, hunting or target	230.7
63	Door closures, automatic, of base metal	228.3
64	Valves, pressure reducing	225.0
65	Garments nes, of materials nes, knit	222.9
66	Safety glass (tempered) for vehicles, aircraft, etc.	218.0
67	Rubber tube, pipe, hose, metal reinforced, no fittings	207.8
68	Nonwovens, man-made fila	207.1
69	Vermouth and other flavoured grape wines—pack <2 l	203.8
70	Wadding, products, manmade fibres, not sanitary items	200.8
71	Sawing machines for stone, ceramics and glass	198.0
72	Uncooked egg pasta not stuffed or prepared	195.7
73	Bars and rods, rectg cross-se	194.0
74	Machine tools to bend, fold, shear or press metal, nes	191.9
75	Drilling or morticing machines for wood, etc.	190.1

(continued)

Table 3.9 (continued)

Rank	Commodity description	Trade balance value (million $)
76	Automatic beverage-vending machines: incorporating heating/ refriger.devices	180.9
77	Containers nes, outer surface of leather	174.4
78	Play, fair-ground equipment, travelling circus, theatre	172.1
79	Rollers, soil preparation, cultivation machinery, nes	172.0
80	Beans, shelled, prepared/preserved, not frozen/vinegar	168.6
81	Calendering or rolling machines, not for metals/glass	167.2
82	Industrial furnace, oven, incinerator non-electric nes	166.4
83	Machinery to impregnate textiles, make linoleum, etc.	164.5
84	Shawls, scarves, etc., of wool or hair, not knit	161.2
85	Non-electric heaters (with fan), parts, of iron/steel	159.5
86	Distilling or rectifying plant	152.2
87	Parts, accessories nes, wood, plastic machine tools	150.1
88	Circular knitting machines, diameter <165 mm	149.7
89	Ingots and other primary forms, stainless steel	148.0
90	Sacks and bags, of paper, having a width > 40 cm	147.7
91	Woven fabric, >85 % combed wool or fine hair, >300 g/m^2	146.3
92	Machines for manufacturing or hot working glass or glassware	144.6
93	Pipe fittings, butt welding of stainless steel	136.8
94	Machinery to prepare, tan, work hides, skins, leather	131.4
95	Mens, boys ensembles, of material nes, not knit	129.5
96	Animal or vegetable fertilizers, in packs >10 kg	127.2
97	Machines for balancing mechanical parts, nes	125.2
98	Corks, crown, of base metal	124.4
99	Watch cases of, or clad with, precious metal	122.8
100	Machinery for preparation of fruits, nuts, vegetables	120.2
101	Built-in jacking systems for garages	118.9
102	Shearing (except punch-shear) machine tools, nes	117.6
103	Ski-boots, cross-country, ski footwear and snowboard boots	116.5
104	Bar/rod, cold formed/finished free cutting steel	115.7
105	Woven cotton nes, >85 % <200 g/m^2, yarn dyed	115.3
106	Bakery ovens, etc., non-electric	113.7
107	Washing, bleaching or dyeing machines (non-domestic)	113.0
108	Medical, surgical or laboratory sterilizers	111.9
109	Tartaric acid	110.6
110	Parts of watch cases	109.1
111	Shawls, scarves, etc., of silk etc., not knit	108.6
112	Woven fabric >85 % artificial filament/strip, dyed, nes	106.0
113	Yarn of carded wool, >85 % wool, not retail	104.2
114	Yarn of carded, wool, <85 % wool, not retail	103.8
115	Bars, rods and profiles, copper alloy nes	101.8

(continued)

Table 3.9 (continued)

Rank	Commodity description	Trade balance value (million $)
116	Footwear, sole leather and uppers textile material	101.2
117	Paper, fine, woodfree, >150 g/m^2, uncoated, nes	100.6
118	Veg fats, oils nes, fractions, not chemically modified	99.8
119	Bar/rod, iron or non-alloy steel, nes	99.1
120	Machinery for producing or preparing textile yarn nes	94.5
121	Dies for drawing or extruding metal	94.1
122	Woven fabric, >85 % carded wool or fine hair, >300 g/m^2	93.3
123	Furnace burners for liquid fuel	91.7
124	Flat not in coils, not further worked than cold-rolled <1>	90.1
125	Tube mills, metal rolling	89.6
126	Other flat, not in coils, not further worked than hot-rolled<4>3	89.2
127	Parts etc., of spinning, combing, drawing machines nes	86.2
128	Finishing agents and dye carriers—textile industry	84.1
129	Automatic vending machines	83.3
130	Electric signal, safety and traffic controller parts	82.8
131	Trees, edible fruit or nut, shrubs and bushes	81.5
132	Parts of machinery for leather industry	81.5
133	Woven fabric, combed wool or hair with natural fibres	81.0
134	Ironing machines and presses including fusing presses	80.3
135	Machinery to reel, fold, cut, pink, etc., textile fabric	79.0
136	Presses, crushers etc., for wine, fruit juice, beverages	78.9
137	Worked calcareous stone nes	77.1
138	Wallpaper and similar wall coverings, nes	74.1
139	Woven fabric of silk, nes	72.4
140	Egg albumin, dried	71.4
141	Woven fabric, >85 % carded wool or fine hair, <300 g/m^2	70.4
142	Chamois (including combination chamois) leather	69.3
143	Clothing accessories nes, of leather or composition	67.2
144	Gravure printing machinery	65.3
145	Cyclan-, cyclen-, cycloterpen-poly-carboxylic acids	64.8
146	Synthetic organic tanning substances	62.9
147	Other automatic vending machines: incorporating heating/refriger. devices	62.6
148	Bovine meat salted, dried or smoked	62.1
149	Womens, girls dresses, of wool or hair, not knit	61.3
150	Knit, crochet fabric of wool or fine animal hair, nes	61.0
151	Womens, girls ensembles, of cotton, not knit	60.0
152	Keys, including blanks for keys, of base metal	59.9
153	Rusks, toasted bread and similar toasted products	59.1
154	Grinding, sanding, polishing machines for wood, etc.	58.5
155	Chicory, fresh or chilled, except witloof	58.3

(continued)

Table 3.9 (continued)

Rank	Commodity description	Trade balance value (million $)
156	Electrodes etc., of base metal or metal carbide, nes	57.0
157	Dairy machinery	56.9
158	Textile fibre preparing machines nes	54.4
159	Pneumatic power engines/motors, except linear acting	54.0
160	Yarn of wool or fine hair, <85 % such fibres, retail	54.0
161	Bars and rods s/steel nfw ho	53.2
162	Woven cotton nes, <85 % + manmade fibre, >200 g, yarn dyed	52.4
163	Fixed power capacitors (50/60 herz circuits)	52.2
164	Wheat meal	51.5
165	Beans nes, prepared or preserved, not frozen/vinegar	51.4
166	Flat rld prd 4 faces <150	48.3
167	Buttons, nes	46.2
168	Corrugated sheets of asbestos, cellulose fibre cement	46.2
169	Womens, girls trousers, shorts, wool or hair, not knit	45.7
170	Camel-back strips for retreading rubber tyres	44.6
171	Expanded metal, i/nas <3 mm wire, <100 cm mesh	44.4
172	Knit or crochet fabric, width <30 cm, >5 % elastomer	43.9
173	Paper, sulphite wrapping, uncoated	43.1
174	Angles, shapes and sections	42.8
175	Milling machines nes, for metal	40.5
176	Hard rubber (e.g. ebonite) in all forms, articles, scrap	39.7
177	Parts and accessories of revolvers or pistols	38.9
178	Copper screws, bolts or nuts except wood screws	37.9
179	Twill weave cotton, <85 % + manmade fibre, >200 g, dyed	36.5
180	Gold, silversmith wares, base clad with precious metal	36.0
181	Reel fed letterpress printers except flexographic	34.4
182	Orthophthalic acid esters, nes	33.7
183	Paper, clay coated, nes	32.1
184	Yarn of wool or fine hair, >85 % such fibres, retail	30.5
185	Yarn >85 % nylon, etc., staple fibre, multiple, not retail	30.1
186	Way-type unit head machines, metal working	28.1
187	Woven fabric <85 % artificial staples, yarn dyed, nes	26.0
188	Couscous	25.7
189	Machinery for leather, skin goods making except sewing	25.2
190	Jewellery, parts, base metal clad with precious metal	24.1
191	Womens, girls suits, of wool or hair, not knit	24.0
192	Endless synchronous bel	21.6
193	Linoleum	20.8
194	Woven fabric of artificial filament, printed, nes	19.9
195	Woven cotton fabric, >200 g/m^2, yarns mixed colours	19.8
196	Woven fabric acrylics + wool or hair, nes	19.4
197	Womens, girls ensembles, of wool or hair, not knit	19.2

(continued)

Table 3.9 (continued)

Rank	Commodity description	Trade balance value (million $)
198	Ties, bow ties and cravats, material nes, not knit	18.4
199	Converters used in metallurgy or metal foundries	17.9
200	Paper, tarred, bituminised or asphalted, nes	17.4
201	Woven cotton nes, >85 % >200 g/m^2, yarn dyed, nes	17.1
202	Mens, boys jackets and blazers, wool or hair, knit	16.8
203	Woven fabric, combed wool or hair + manmade filament	16.4
204	Ketone-alcohols nes; ket	16.4
205	Blankets (non-electric) and travelling rug, wool or hair	15.8
206	Parts of presses, crushers for wine, fruit juice etc.	15.7
207	Woven fabric >85 % nylon, polyamide, printed, nes	15.3
208	Woven fabric >85 % artif filament/strip, yarn dyed, nes	15.2
209	Woven fabric of artificial filament, yarn dyed, nes	14.9
210	Vermouth and other flavoured grape wines—pack >2 l	14.7
211	Twill weave cotton, >85 % >200 g/m^2, bleached	14.2
212	Pocket-watch, precious-metal case, non-battery	13.4
213	Homogenized vegetable preparations, not frozen/vinegar	13.3
214	Woven fabric, carded wool or hair with natural fibre	13.2
215	Textile fibre carding machines	13.1
216	Woven fabric <85 % artificial staple + cotton, yarn dyed	13.1
217	Super-heated water boilers	13.0
218	Sulphites of metals other than sodium	11.7
219	Other Ski-boots, cross-country, ski footwear and snowboard boots	11.5
220	Woven fabric of noil silk	11.3
221	Twill weave cotton, <85 % + manmade fibre, <200 g/m^2 dyed	10.9
222	Woven fabric >85 % artif filament/strip, printed, nes	10.7
223	Flax yarn multiple (folded) or cabled	9.7
224	Woven fabric, carded wool or hair + manmade filament	9.0
225	Woven fabric of hi-ten filament yarns of viscose rayon	8.8
226	Woven fabric >85 % nylon, polyamide, yarn dyed, nes	7.7
227	Woven fabric <85 % artif staple + wool or hair, yarn dyed	6.9
228	Loop pile knit/crochet fabric, natural yarn not cotton	6.4
229	Woven fabric <85 % artificial staple + wool/hair, printed	6.0
230	Citrus based jams jellies marmalade, etc.	5.8
231	Plain weave cotton, <85 % + manmade fibre, >200 g, yarn dyed	3.2
232	Maps and charts, printed, in book form	2.6
233	Ephedrines and their salts	2.4
234	Carbon tetrachloride	2.4
235	Monochrome cathode-ray picture tubes, monitors	0.8

Index of Italy's competitive excellence in world trade: Fortis-Corradini's Index (case study of a total sample of 5,117 products that comprises the International Trade)
Source Compiled by Fondazione Edison on data from Istat, Eurostat and UN Comtrade

Table 3.10 Products in which Italy holds second place in the world trade balance: year 2012

Rank	Commodity description	Trade balance value (million $)
1	Taps, cocks, valves and similar appliances, nes	4,932.7
2	Grape wines nes, fortified wine or must, pack <2 l	4,559.6
3	Furniture, wooden, nes	1,877.9
4	Furniture parts nes	1,870.7
5	Articles of iron or steel, nes	1,857.0
6	Wheeled tractors nes	1,561.9
7	Unglazed ceramic flags, tiles >7 cm wide	1,531.8
8	Parts of gas turbine engines except turbo-jet/prop	1,438.8
9	Machinery to fill, close, aerate, etc., bottle, container	1,431.8
10	Electric conductors, 80–1,000 V, no connectors	1,297.6
11	Articles of aluminium, nes	1,211.2
12	Parts of machines and mechanical appliances nes	1,164.5
13	Motorboats, other than outboard motorboats	1,117.2
14	Coffee, roasted, not decaffeinated	1,055.5
15	Parts of vacuum pumps, compressors, fans, blowers, hoods	819.4
16	Communion wafers, rice paper, bakers wares nes	756.3
17	Grapes, fresh	723.6
18	Heat exchange units, non-domestic, non-electric	682.4
19	Machinery for treatment by temperature change nes	666.2
20	Grape wines, sparkling	639.6
21	Tomatoes nes, prepared or preserved, not in vinegar	609.4
22	Pullovers, cardigans etc., of wool or hair, knit	592.0
23	Worked marble, travertine and alabaster	583.9
24	Rubber or plastic working machines, nes	535.5
25	Articles for pocket or handbag, leather outer surface	525.1
26	Cooking appliances for gas fuel, etc., iron or steel	518.8
27	Parts of turbo-jet or turbo-propeller engines	481.6
28	Motorcycles, spark ignition engine of >800 cc	471.3
29	Womens, girls garments nes, material nes, not knit	466.2
30	Pipe and tube fittings, of copper alloys	448.0
31	Parts of refrigerating or freezing equipment	429.4
32	Mineral and aerated waters not sweetened or flavoured	428.8
33	Fittings, pipe or tube, iron or steel, nes	427.3
34	Pipe flanges, iron or steel except stainless/cast	409.6
35	Grape wines, alcoholic grape must nes	409.4
36	Mens, boys suits, of wool or hair, not knit	402.8
37	Parts of taps, cocks, valves or similar appliances	395.1
38	Liqueurs and cordials	392.3
39	Electric coffee or tea makers, domestic	391.3

(continued)

Table 3.10 (continued)

Rank	Commodity description	Trade balance value (million $)
40	Num controlled machine tools to bend, fold, etc., metal	391.3
41	Panty hose etc., synthetic fibre yarn <67 dtex/sy, knit	377.7
42	Steam or sand blasting machines	377.4
43	Stoppers, caps, lids, packing items nes of base metal	377.1
44	Central heating boilers nes	375.6
45	Sausages, similar products of meat, meat offal and blood	375.6
46	Ventilating hoods having a maximum width < 120 cm	368.1
47	Kiwifruit, fresh	357.8
48	Parts of lifts, skip hoist or escalators	350.6
49	Reservoirs/tanks/vats/etc., iron/steel capacity >300 l	342.5
50	Paper, household, sanitary, width > 36 cm	317.3
51	Cut or sawn slabs of marble, travertine or alabaster	314.5
52	Parts of balloons, dirigibles, spacecraft	312.3
53	Seats with metal frames, upholstered nes	305.2
54	Air conditioners nes, without refrigerating unit	303.4
55	Plants live, mushroom sp	299.0
56	Water filtering or purifying machinery or apparatus	294.9
57	Peaches, nectarines, fresh	289.3
58	Table, kitchen and household	287.4
59	Butt weld fittings, iron/steel except stainless/cast	287.1
60	Semi-finished products of alloy steel except stainless	284.1
61	Soles and heels for footwear, of rubber or plastic	278.5
62	Furniture of cane, materials nes	269.9
63	Vegetables, fresh or chilled nes	263.7
64	Poultry-keeping machinery, nes	259.8
65	Safety glass (laminated) for vehicles, aircraft, etc.	256.0
66	Seats nes	251.8
67	Containers for compressed/liquefied gas, iron or steel	237.2
68	Metal treating machines, electric wire coil-winders	234.8
69	Toilet paper	232.9
70	Parts of industrial or laboratory furnaces/ovens	227.4
71	Eye make-up preparations	226.5
72	Trays, dishes, plates, cups and the like, of paper	225.3
73	Polyester based paints, varnishes, non-aqueous medium	221.4
74	Garments made up of textile felts and nonwoven fabric	221.4
75	Industrial machinery nes for food, drink preparation	221.3
76	Parts of central heating boiler nes	211.4
77	Ties, bow ties and cravats, of silk, not knit	210.6
78	Earth moving/road making equipment, self-propelled nes	205.7

(continued)

Table 3.10 (continued)

Rank	Commodity description	Trade balance value (million $)
79	Furniture, plastic, nes	204.7
80	Machines, confectionery, cocoa, chocolate manufacture	199.2
81	Plastic household, toilet articles not table, kitchen	198.2
82	Machine tools to forge, stamp, hammer or press metals	192.1
83	Isolating and make-and-break switches, voltage >1 kV	191.4
84	Paper, glassine, glazed transparent or translucent nes	191.0
85	Machines to mix, knead, crush, grind, etc., nes	190.7
86	Casting machines used in metallurgy, foundries	189.0
87	Olive oil, fractions, refined, not chemically modified	187.7
88	Anti-oxidisers and stabilizers for rubber or plastics	185.9
89	Electric instant, storage and immersion water heaters	183.9
90	Blow moulding machines for rubber or plastic	183.1
91	Parts nes of machinery for making pulp, paper or board	178.0
92	Sacks and bags, of paper, nes, including cones	174.9
93	Moulds, injection or compression, for metals/carbides	174.7
94	Parts, accessories nes, stone, ceramic machine tools	167.6
95	Draw-benches for bars, tubes, profiles wire etc.	166.2
96	Office furniture, metal, nes	166.1
97	Clutches, shaft couplings, universal joints	165.4
98	Doors, frames and thresholds, of wood	165.0
99	Womens, girls overcoats etc., of wool or hair not knit	163.2
100	Marble and travertine in blocks etc.	155.7
101	Tower cranes	153.9
102	Tube, pipe or hose, rigid, of polyethylene	152.7
103	Sprays/powder dispersing machines except agricultural	148.5
104	Nonwovens, man-made fila	147.9
105	Polyvinyl chloride nes, plasticised in primary forms	147.4
106	Trunks, suit-cases/etc., outer surface leather	146.2
107	Flexographic printing machinery	145.3
108	Furnace burners for solid, gas or combination fuel	144.6
109	Tiles and flagstones of cement, concrete, artificial stone	143.5
110	Test benches for measuring or checking equipment	143.4
111	Parts of furnace burners, associated equipment	142.0
112	Auxiliary plant for steam/vapour generating boilers	141.9
113	Cans, iron or steel, capacity <50 l nes	140.7
114	Mens, boys jackets and blazers, wool or hair, not knit	139.0
115	Parts for domestic non-electric thermic appliances	137.5
116	Multi-station transfer machines for working metal	135.4
117	Cans, iron/steel, capacity <50 l closed by crimp/solder	135.4

(continued)

Table 3.10 (continued)

Rank	Commodity description	Trade balance value (million $)
118	Vapour generating boilers nes, hybrid boilers	132.8
119	Scarifiers, cultivators, weeders and hoes	132.7
120	Cheese, blue-veined	130.7
121	Motorcycles, spark ignition engine of <50 cc	130.0
122	Children's picture, drawing or colouring books	129.0
123	Parts of industrial/etc., electric furnaces/ovens nes	127.5
124	Womens, girls dresses, of artificial fibres, not knit	126.9
125	Physical exercise, gymnasium and athletics equipment	119.8
126	Mountings, fittings, of base metal, nes	118.9
127	Parts of equipment for metallurgy, foundries	118.5
128	Presses for particle, fibre board, etc., manufacture	117.6
129	Compounded unvulcanised rubber in plate, sheet, strip	115.8
130	Registers, account books, note books, diaries etc.	113.7
131	Tube, pipe or hose, rigid, of polyvinyl chloride	112.6
132	AC generators, of an output <75 kVA	108.8
133	Footwear, uppers of leather with metal toe-cap, nes	108.4
134	Pipe fittings of non-malleable cast iron	108.1
135	Electric table, desk, bedside and floor lamps	105.5
136	Multi-purpose machines for wood etc., work	105.4
137	Powders, for skin care and make-up	102.9
138	Hot, or combination hot-cold metal rolling mills	101.6
139	Circular saw blades, working part of steel	97.8
140	Lucerne (alfalfa) meal and pellets	96.8
141	Non-refractory mortars a	96.4
142	Glues or adhesives, prepared nes, package >1 kg	95.3
143	Machinery for cleaning/drying bottles/containers nes	94.3
144	Pneumatic elevators and conveyors	94.2
145	Bicycle saddles	93.8
146	Womens, girls suits, of material nes, not knit	93.2
147	Moulds for metal or metal carbides, nes	90.7
148	Panty hose etc., synthetic fibre yarn >67 dtex/sy, knit	90.4
149	Teleferiques, chair-lifts, ski-lifts, etc.	89.4
150	Machines to process animal or fixed veg fats or oils	88.8
151	Grindstones etc., of synthetic or natural diamond	88.7
152	Cutting machines for paper pulp, paper or paperboard	85.0
153	Machinery for making or repairing footwear	84.5
154	Tank, cask or container, iron/steel, capacity 50–300 l	84.1
155	Handbags, of vulcanised fibre or paperboard	83.0
156	Paper, kraft, >95 % chem pulp, <150 g, bleached, clay coat	82.6

(continued)

Table 3.10 (continued)

Rank	Commodity description	Trade balance value (million $)
157	Parts and accessories equally suitable for use with machines	82.2
158	Plate, sheet, strip of vulcanised non-cellular rubber	82.1
159	Mens, boys trousers and shorts, wool or hair, not knit	81.4
160	Pipe fittings of malleable iron or steel, cast	81.4
161	Bars, rods and profiles, nickel alloy	80.8
162	Winches or capstans nes powered by electric motor	80.6
163	Threshing machinery nes	77.1
164	Safety headgear	76.2
165	Cartridges, shotgun	76.1
166	Parts, accessories for gas, liquid, electricity meters	74.6
167	Womens, girls jackets, blazers, wool or hair, not knit	74.1
168	Mens, boys overcoats of wool or hair, not knit	73.5
169	Monofilament (>1 mm), rods, not ethylene or vinyl polymer	73.4
170	Drilling machines for metal, except num controlled	72.4
171	Worked monumental or building stone nes	69.1
172	Glass cubes, mosaic tiles	67.4
173	Circular knitting machines, diameter >165 mm	66.7
174	Woven twill >85 % polyester + cotton, >170 g/m^2 dyed	66.4
175	Woven fabric, >85 % flax, except unbleached or bleached	66.2
176	Mens, boys suits, of material nes, not knit	64.8
177	Woven fabric >85 % textured polyester, printed, nes	64.4
178	Parts of garments, clothing accessories nes, not knit	63.8
179	Brushes, parts, nes	63.6
180	Sinks and wash basins, stainless steel	62.9
181	Dry-cleaning machines	61.5
182	Plain weave cotton, >85 % 100–200 g/m^2, yarn dyed	60.4
183	Mobile lifting frames on tyres, straddle carriers	59.9
184	Angles, shps and sec from fla	59.9
185	Threaded fittings, iron or steel except stainless/cast	59.9
186	Marble granules, chipping and powder	59.5
187	Cold metal rolling mills	59.0
188	Bridges and bridge sections, iron or steel	58.5
189	Parts of automatic goods-vending machinery	58.4
190	Electrical insulators of glass	58.4
191	Dryers for agricultural products	57.8
192	Grinding machines nes, accurate to 0.01 mm	57.4
193	Machines to extrude, draw, cut manmade textile fibres	57.3
194	Woven fabric of flax, <85 % flax, except unbl/bleached	57.2
195	Aluminium sanitary ware, parts thereof	57.1

(continued)

Table 3.10 (continued)

Rank	Commodity description	Trade balance value (million $)
196	Flat, not further worked than hot-rolled: plated or coated with aluminium <4.75	56.1
197	Household, hospital and sanitary supplies, of paper, nes	55.9
198	Sulphonated hydrocarbons, salts and ethyl esters	55.3
199	Coal or rock cutters, not self-propelled	54.8
200	Flat, in coils, not further worked than cold-rolled <3 mm	54.4
201	Woven fabric of artificial filament, dyed, nes	54.0
202	Transformers electric, power capacity 16–500 KVA	54.0
203	Pumps fitted, designed to have, a measuring device nes	53.9
204	Vitrifiable enamels, glazes etc., for ceramics and glass	53.6
205	Rubber tube, pipe or hose, reinforced nes, no fittings	53.2
206	Parts of machines for treating textile fabrics	50.5
207	Needleloom felt and stitch-bonded fibre fabric	49.5
208	Machinery for finishing paper or paperboard	47.4
209	Twill weave cotton, >85 % <200 g/m^2, dyed	47.2
210	Mens, boys garments nes, of material nes, not knit	47.2
211	Dentists, barbers or similar chairs and parts	47.1
212	Waterproof footwear (Wellington) no toe cap, over ankle	45.7
213	Fixed wing aircraft, unladen weight <2,000 kg	45.4
214	Asphalt or similar material articles not in rolls	44.1
215	Refractory bricks etc., nes	43.8
216	Mixing, kneading machines for minerals except bitumen	43.3
217	Planing and milling machines for wood, cork, etc.	42.1
218	Womens, girls skirts, of material nes, not knit	42.1
219	Mats of glass fibres	41.6
220	Machines for working metal wire	41.4
221	Pocket-watch, base-metal case, non-battery	41.2
222	Surface grinding machines accurate to 0.01 mm nes	40.2
223	Pears, otherwise prepared or preserved	40.0
224	Bar/rod of silico-manganese steel not in coils	39.4
225	Essential oils of citrus fruits, nes	39.3
226	Hams and shoulders, swine, salted, dried or smoked	39.2
227	Threading or tapping machines nes for metal	39.1
228	Shawls, scarves, etc., of artificial fibres, not knit	38.7
229	Tailors dummies, display dummies and automata	38.4
230	Rubberised woven textile fabric, except adhesive tape	37.8
231	Sheet/film not cellular/reinf cellulose acetate	35.7
232	Hot rolled stainless steel flat, w >600 mm, t 3–4.75 mm	35.7
233	Silver wares, silver ware plated with precious metal	34.6

(continued)

Table 3.10 (continued)

Rank	Commodity description	Trade balance value (million $)
234	Mixtures of edible nuts, dried and preserved fruits	34.6
235	Saw blades nes, including stone cutting, etc.	34.0
236	Grape must, unfermented, except as fruit juice	33.2
237	Womens, girls dresses, of wool or hair, knit	33.0
238	Articles of copper, cast/moulded/stamped, nfw	33.0
239	Parts/access, machines for preparing manmade textiles	32.4
240	Woven cotton nes, >85 % <200 g/m^2, printed	31.9
241	Tyre cord fabric of viscose rayon	31.8
242	Olive oil, fractions, blends, not chemically modified	31.3
243	Filtering/purifying machinery/apparatus for beverages	31.2
244	Essential oils of lemon	31.2
245	Sodium sulphites	30.8
246	Shawls, scarves, veils etc., textile material, knit	30.7
247	Felt, impregnated, coated, covered, or laminated	29.0
248	Liquid elevators	28.6
249	Parts, access nes, knitting, embroidery, lace machines	28.5
250	Office binder/file fittings, of base metal	28.4
251	Statuettes and other ornaments, base metal, unplated	28.4
252	Womens, girls suits, of cotton, not knit	28.2
253	Mens, boys anoraks etc., of wool or hair, not knit	27.7
254	Yarn of combed fine animal hair, not retail	27.5
255	Articles of furskin except clothing and accessories	27.4
256	Woven fabric, combed wool or hair + manmade staple fib	27.1
257	Womens, girls overcoats, etc., of wool or hair, knit	26.9
258	Inflatable pleasure craft	26.9
259	Washers, spring or lock, iron or steel	26.0
260	Stone mosaic tiles, artificial coloured chips etc.	26.0
261	Woven fabric synthetic filament, yarn dyed, nes	26.0
262	Welding machinery not gas-operated	25.8
263	Ingot moulds and ladles used in metallurgy/foundries	25.6
264	Woven cotton fabric, >200 g/m^2, dyed, nes	25.3
265	Plate/sheet/strip, copper-zinc alloy, flat, t > 0.15 mm	25.3
266	Wrist-watch, precious metal, battery, other	24.9
267	Sections, T, i/nas, nfw hot-roll/drawn/extruded <80 mm	24.9
268	Oleic, linoleic or linolenic acids, salts and esters	24.7
269	Yarn nylon/polyamide filament multiple nes, not retail	24.2
270	Textile yarn/strip, rubber, plastic impregnated/coated	24.0
271	Knit or crochet fabric of other materials, nes	24.0
272	Plain weave cotton, >85 % <100 g/m^2, yarn dyed	22.9
273	Machines for moulding articles in pulp, paper, board	22.5

(continued)

Table 3.10 (continued)

Rank	Commodity description	Trade balance value (million $)
274	Chain, iron or steel, nes	21.9
275	Castors of base metal	21.3
276	Womens, girls skirts, of wool or hair, not knit	20.8
277	Yarn of combed wool, <85 % wool, not retail	20.5
278	Cinematograph film, exposed and developed, width >35 mm	20.4
279	Composition leather, in slabs, sheets or strip	19.5
280	Cooking appliances for solid fuel, iron or steel	19.5
281	Womens, girls panties, bathrobes, etc., nes not knit	19.0
282	Womens, girls jackets and blazers, of wool or hair, knit	18.7
283	Woven fabric <85 % artificial staples, dyed, nes	18.4
284	Hosiery nes, of wool or fine animal hair, knit	18.4
285	Parts of machinery for preparing tobacco	18.3
286	Parts nes for rolling machines, except metals, glass	18.3
287	Tubes and pipe, aluminium, not alloyed	18.2
288	Net/fencing, plastic-coated i/nas <3 mm wire, <100 cm mesh	18.1
289	Machines to assemble electric lamps, glass envelope	18.1
290	Yarn of acrylic staple fibres and cotton, not retail	17.8
291	Thread rolling machines for working metal, etc.	17.7
292	Hats, plaited or made by assembling strips of material	17.5
293	Polyvinyl acetate, in aqueous dispersion	17.1
294	Hot rolled stainless steel sheet, w <600 mm, t >4.75 mm	16.9
295	Woven cotton fabric, >200 g/m^2, dyed, nes	16.9
296	Cellulose acetates, plasticised, in primary forms	16.5
297	Air gun pellets, parts of shotgun cartridges	16.5
298	Snow-ski equipment except skis/bindings	16.4
299	Hand-operated appliances, food preparation, <10 kg	16.2
300	Carboxylic acids (aldehyde, ketone function), derivs	15.7
301	Lathes nes for removing metal	15.0
302	Womens, girls anoraks etc., of wool or hair, not knit	14.9
303	Fittings, pipe and tube, nickel	14.6
304	Twill weave cotton, >85 % <200 g/m^2, yarn dyed	14.4
305	Cereal flour except wheat, meslin, rye, maize, rice	14.3
306	Woven warp pile cotton, cut, except terry, w > 30 cm	14.2
307	Engines, spark-ignition reciprocating, <50 cc	14.0
308	Mens, boys ensembles, of material nes, knit	12.8
309	Calcined dolomite	12.7
310	Headgear nes, of other materials	12.6
311	Woven cotton fabric, <200 g/m^2, yarns mixed colours	12.2
312	Rubber waste, parings and scrap (except hard rubber)	11.8

(continued)

Table 3.10 (continued)

Rank	Commodity description	Trade balance value (million $)
313	Label, badge, etc., non-embroidered textile, not woven	11.8
314	Other flat not in coils, not further worked than cold-rolled <3 mm	11.7
315	Yarn >85 % nylon, etc., staple fibres, single, not retail	11.6
316	Molybdenum, articles thereof nes	11.5
317	Woven fabric acrylic or modacrylic staple fibres, nes	11.4
318	Yarn <85 % synthetic staple fibres, retail not sewing	11.3
319	Yarn of viscose rayon, multiple, nes, not retail	11.1
320	Smoking pipes and pipe bowls	11.0
321	Amino-aldehydes, ketones and quinones, salts thereof	10.9
322	Sheep or lamb skins, raw, except pickled, no wool	10.6
323	Screw hooks and screw rings of iron or steel	10.4
324	Other Automatic beverage-vending machines	10.3
325	Woven fabric acrylics + manmade filament, nes	10.2
326	Animal carving material, articles, nes	10.2
327	Apples, dried	10.0
328	Woven cotton fabric, >200 g/m^2, printed, nes	9.9
329	Cotton yarn <85 % multiple uncombed >714, not ret., nes	9.0
330	Womens, girls trousers and shorts, of wool hair, knit	8.9
331	Woven fabric materials nes, <30 cm wide	8.6
332	Woven fabric <85 % artificial staple + wool or hair, dyed	8.3
333	Woven fabric >85 % synth nes + cotton, >170 g/m^2 yarn dye	8.3
334	Twill weave cotton, >85 % <200 g/m^2, bleached	8.1
335	Staple fibres nylon, polyamides, carded or combed	7.9
336	App and equip for ph labor	7.6
337	Hexamethylenediamine, its salts	7.4
338	Parts of lighters, other than flints/wicks	7.1
339	Woven cotton nes, <85 % + manmade fibre, <200 g/m^2 print	7.0
340	Woven fabric of other vegetable textile fibre, paper	6.8
341	Yarn of other synthetic staple with wool or hair, nes	6.8
342	Albumins nes, albuminates and other albumin derivatives	6.7
343	Handkerchiefs, of material nes, not knit	6.7
344	Paper, sensitising base stock	6.7
345	Thiuram mono-, di- or tetra-sulphides	6.7
346	Tomato juice not fermented or spirited	6.6
347	Pile knit, crochet fabric, natural yarn not cotton nes	6.5
348	Clocks, nes, non-electric	6.1
349	Cutlery sets plated with precious metal	5.5
350	Womens, girls blouses and shirts, wool or hair, not knit	5.5
351	Shuttle type looms nes for weaving fabric >30 cm wide	5.5

(continued)

Table 3.10 (continued)

Rank	Commodity description	Trade balance value (million $)
352	Rubber thread and cord, textile covered	5.4
353	Twill cotton not denim <85 % + manmade fibre, >200 g y-dyed	5.2
354	Tracing cloth, painting canvas, stiffened textile nes	5.0
355	Alkaloids, derivs, without antibiotics, hormones, bulk	4.6
356	Woven fabric <85 % artificial staples, printed, nes	4.3
357	Yarn of artificial fibres except sewing thread, retail	4.0
358	Yarn >85 % other synth staple fibres, single not retail	4.0
359	Womens, girls skirts, of wool or hair, knit	3.6
360	Woven fabric of flax, <85 % flax, unbleached/bleached	3.1
361	Harpsichords, keyboard stringed instruments nes	2.8
362	Umbrella frames	2.6
363	Cotton yarn <85 % multiple combed >714, not retail, nes	2.6
364	Textile fibre combing machines	2.5
365	Woven fabric incorporating metal threads, nes	2.4
366	Cotton yarn >85 % single uncombed <125 dtex, not retail	2.3
367	Chenille cotton fabric, width > 30 cm	2.1
368	Mens, boys trousers and shorts, of wool or hair, knit	2.0
369	Cotton yarn <85 % multiple combed <125 dtex, not retail	1.9
370	Photo plates and film, exposed and developed, for offset	1.8
371	Cotton yarn <85 % multiple combed 714–232 dtex, not ret.	1.6
372	Yarn, cellulose acetate, multiple, nes, not retail	1.6
373	Cotton yarn <85 % single uncombed <125 dtex, not retail	0.9
374	Yarn, viscose rayon, single >120 turn/m nes, not retail	0.8
375	Ergotamine (INN) and it	0.6
376	Anti-knock preparations based on lead compounds	0.2

Index of Italy's competitive excellence in world trade; Fortis-Corradini's Index (case study of a total sample of 5,117 products that comprises the International Trade)
Source Compiled by Fondazione Edison on data from Istat, Eurostat and UN Comtrade

Table 3.11 Products in which Italy holds third place in the world trade balance: year 2012

Rank	Commodity description	Trade balance value (million $)
1	Jewellery and parts of precious metal except silver	4,281.9
2	Motor vehicle parts nes	3,786.9
3	Glazed ceramic flags, tiles wider than 7 cm	2,311.0
4	Machines and mechanical appliances nes	2,199.7
5	Plastic articles nes	1,417.0
6	Gearing, ball screws, speed changers, torque converter	1,369.0
7	Air or gas compressors, hoods	1,259.2
8	Seats with wooden frames, upholstered nes	1,217.3
9	Brake system parts except linings for motor vehicles	1,147.0
10	Drive axles with differential for motor vehicles	1,012.6
11	Structures and parts of structures, iron or steel, nes	938.7
12	Footwear, sole rubber, plastics uppers of leather, nes	918.4
13	Containing indentations, ribs, grooves, etc produced during rolling process	862.1
14	Bolts/screws nes, with/without nut/washer, iron/steel	838.5
15	Other flat, not in coils, not further worked than hot-rolled <10 mm	777.9
16	Kitchen furniture, wooden, nes	742.7
17	Automatic washing machines, of a dry capacity <10 kg	704.4
18	Plastic sheet, film, foil or strip, nes	690.3
19	Special purpose motor vehicles nes	682.9
20	Chandeliers, other electric ceiling or wall lights	637.9
21	Electric cooking, grilling and roasting equipment nes	591.1
22	Boots, sole rubber or plastic upper leather, nes	515.3
23	Lifts and skip hoists	486.6
24	Worked granite	470.9
25	Pipes, line, iron or steel, for oil or gas pipelines	453.3
26	Parts for mineral sort, screen, mix, etc., machines	430.7
27	Wheels including parts/accessories for motor vehicles	429.3
28	Boring or sinking machinery nes, self-propelled	423.1
29	Articles of apparel of leather or composition leather	412.3
30	Valves for oleohydraulic or pneumatic transmissions	407.3
31	Articles of vulcanised rubber nes, except hard rubber	382.2
32	Refrigerating or freezing equipment nes	381.8
33	Furniture, metal, nes	366.7
34	Extruders for working rubber or plastic	362.8
35	Trade advertising material, catalogues etc.	352.2
36	Sheet/film not cellular/reinf polymers of ethylene	329.8
37	Polyurethanes in primary forms	324.7
38	Parts, laboratory/industrial heating/cooling machinery	323.3
39	Lifting, handling, loading or unloading machinery nes	292.9

(continued)

Table 3.11 (continued)

Rank	Commodity description	Trade balance value (million $)
40	Electric lamps, lighting fittings, nes	292.9
41	Air conditioners nes, with refrigerating unit	292.7
42	Handbags with outer surface plastics, textile material	292.6
43	Sauces nes, mixed condiments, mixed seasoning	282.5
44	Parts of machines for working rubber or plastic	276.7
45	Office furniture, wooden, nes	275.4
46	Textile yarn winding or reeling machines	255.6
47	Sheet/film not cellular/reinf flexible vinyl polymer	251.1
48	Articles of apparel and clothing accessories of furskin	232.0
49	Parts for sprays and powder dispersers	223.2
50	Steel pipes nes, longitudinally welded diameter >406 mm	222.3
51	Plastic stoppers, lids, caps and other closures	221.0
52	Machines for public works, building etc., nes	220.8
53	Industrial machine parts, food, drink preparation	219.3
54	Frames and mountings for spectacles etc., except plastic	208.3
55	Vacuum pumps	207.1
56	Continuous action elevators or conveyors for goods nes	205.8
57	Plastic tube, pipe or hose, flexible, nes	205.1
58	Concrete or mortar mixing machines	200.7
59	Mowers, powered, lawn, with horizontal cutting device	196.4
60	Reciprocating positive displacement pumps nes	190.0
61	Fittings for plastic tube, pipe or hose	187.6
62	Parts of machines for making paper or paperboard	186.7
63	Grape juice or must not fermented or spirited	181.3
64	Electric fans, motor >125 W	175.8
65	Stranded steel wire/cable/etc., no electric insulation	162.6
66	Clothing accessories nes, textile material, not knit	160.6
67	Gaskets, washers and other seals of vulcanised rubber	160.4
68	Printing machinery nes	157.8
69	Scent sprays and similar toilet sprays, parts	156.5
70	Penicillins and streptomycins, derivs, in dosage	151.0
71	Industrial electric resistance heated furnaces and ovens	149.8
72	Tomato ketchup and other tomato sauces	147.3
73	Mountings, fittings, of base metal, for buildings, nes	143.0
74	Machines to mill or work cereals or dried legumes nes	142.2
75	Wallpaper, coated or covered with a decorated plastic	142.0
76	Machine-tools (metal, carbides) not metal removal, nes	139.7
77	Ceramic bathroom kitchen sanitary items not porcelain	139.1
78	Marble and travertine, crude or roughly trimmed	137.3
79	Flanges, stainless steel	137.2
80	Polymer based paints and varnishes nes, aqueous medium	132.9

(continued)

Table 3.11 (continued)

Rank	Commodity description	Trade balance value (million $)
81	Ice cream and other edible ice	126.6
82	Non-domestic, non-electric dryers nes	122.8
83	Moulds for rubber or plastic, nes	120.2
84	Shuttle-less looms for weaving fabric >30 cm wide	120.1
85	Foil, aluminium, backed, <0.2 mm thick except backing	119.4
86	Machinery for the preparation of meat and poultry	117.5
87	Plastic articles for goods conveyance or packing nes	116.8
88	Sawing machines for working wood, cork, etc.	116.1
89	Seeders, planters and transplanters	115.1
90	Flywheels and pulleys including pulley blocks	113.3
91	Rotary positive displacement pumps nes	110.2
92	Nonwovens, man-made filaments < 150 > 70	109.1
93	Paper, weighing 225 g/m^2 or more, uncoated, nes	107.5
94	Womens, girls dresses, of cotton, not knit	107.3
95	Metal sawing or cutting-off machines	106.4
96	Wire plated or coated with other base metals	100.6
97	Lettuce, fresh or chilled except cabbage lettuce	99.7
98	Non-automatic electric plasma and other arc welders	99.7
99	Transformers electric, power capacity >500 KVA, nes	98.9
100	Numerically controlled punch, notch, punch-shear machines	96.7
101	Furnaces/ovens non-electric for ores/pyrites/metals	95.3
102	Parts of household or laundry-type washing machines	93.4
103	Pig and poultry fat, unrendered	93.2
104	Paper, filter, cut to size or shape	91.9
105	Parts of electric accumulators, including separators	91.4
106	Wire of alloy steel nes	91.3
107	Parts and accessories for photographic cameras	90.8
108	Grill netting and fencing, welded, i/nas <3 mm wire, <100 cm mesh	90.6
109	Womens, girls blouses and shirts, of silk, not knit	89.2
110	Mobile concrete mixers	89.0
111	Table/kitchen articles, parts, of iron or steel, nes	88.4
112	Fluoro-polymers nes in primary forms	88.0
113	Hydraulic presses for working metal	87.5
114	Rubber tube, pipe, hose textile-reinforced no fittings	86.3
115	Seats with metal frames, nes	85.7
116	Tube/pipe/hollow profile, iron/steel, riveted/open seam	83.6
117	Sound signalling equipment	80.6
118	Glucose, glucose syrup <20 % fructose	80.3
119	Parts of hoists and winches	80.3
120	Citrus juice nes (one fruit) not fermented or spirited	79.0
121	Winches or capstans nes	77.8

(continued)

Table 3.11 (continued)

Rank	Commodity description	Trade balance value (million $)
122	Oral and dental hygiene preparations, except dentifrices	76.5
123	Revolvers and pistols	74.0
124	Table/kitchen articles, parts, stainless steel	73.8
125	Shawls, scarves, etc., of material nes, not knit	72.9
126	Endless synchroneous bel	71.5
127	Rope or cable-making machines	70.2
128	Womens, girls dresses, of artificial fibres, knit	69.7
129	Swine hams and cuts thereof, prepared or preserved	69.5
130	Electrical insulators, except glass/ceramics	69.4
131	Hosiery nes, synthetic fibres, knit	68.6
132	Orange juice, frozen, not fermented or spirited	68.2
133	Stainless steel bar or rod nes	67.0
134	Water-skis, surf-boards, other watersport equipment	66.6
135	Baby carriages and parts thereof	66.6
136	Hot rolled i/nas, w <600 mm, t <3 mm myp275/t >3 mm myp355	66.5
137	Containers, packing, nes of paper	65.4
138	Moulds for mineral materials	64.9
139	Universal AC/DC motors of an output <37.5 W	64.8
140	Hay etc., mowers and cutter bars, tractor mounting	64.7
141	Machines for cleaning, sorting, grading eggs/fruit/etc.	63.7
142	Concrete pumps	63.5
143	Cranes or derricks nes	62.4
144	Toluidines, derivatives, salts thereof	62.3
145	Parts and accessories of looms, auxiliary machinery nes	60.8
146	Womens full, kneelength hosiery, yarn <67 dtex/sy, knit	60.0
147	Watermelons, fresh	59.7
148	Edible brassicas nes, fresh or chilled	59.5
149	Springs, iron or steel, except helical/leaf	58.6
150	Beetroot, salsify, celeriac, radishes etc., fresh, chilled	57.4
151	Air cond used in vehicle	56.7
152	Maize (corn) groats or meal	56.1
153	Rubber tube, pipe or hose not reinforced, no fittings	55.7
154	Polyvinyl chloride nes, not plasticised, primary form	55.5
155	Womens, girls jackets and blazers, material nes, not knit	55.5
156	Cut or sawn slabs of granite	55.1
157	Equipment nes for sports, swimming and paddling pools	54.9
158	Plate, sheet, strip of vulcanised cellular rubber	54.7
159	Parts of poultry-keeping machinery	54.5
160	Rubber or plastic vacuum moulders, thermoformers	53.9
161	Foliage, branches, for bouquets, etc.,—fresh	53.9

(continued)

Table 3.11 (continued)

Rank	Commodity description	Trade balance value (million $)
162	Abrasive powder or grain on a base of other material	52.6
163	Palmitic acid, stearic acid, their salts and esters	52.4
164	Dryers for wood, paper pulp, paper or paperboard	49.7
165	Fruits, nuts, fruit-peel, etc., preserved by sugar	49.1
166	Electrical lamp-holders, for <1,000 V	47.9
167	Industrial electric induction, dielectric furnace/oven	47.4
168	Carboxylic acids (alcohol function only), derivatives	47.2
169	Numerically controlled metal working drill machines	46.3
170	Electric hair dryers	45.8
171	Grill, netting, fencing, ir	45.7
172	"Long pile" knit or crochet textile fabric	44.5
173	Chain, articulated link, iron or steel, except roller	44.2
174	Inorganic tanning and pre-tanning preparations	43.9
175	Textile yarn spinning machines	42.9
176	Transmission or conveyor belts or belting of textile	41.7
177	Angles, shapes and sections, stainless steel	39.8
178	Vinyl acetate copolymers in aqueous dispersion	39.7
179	Bending or assembling machines for wood, etc.	37.7
180	Parts grain, seed, dry legumes processing equipment	37.6
181	Womens, girls overcoats etc., manmade fibre, not knit	37.5
182	Mens, boys ensembles, of cotton, not knit	37.4
183	Threaded elbows, bends and sleeves of stainless steel	37.1
184	Screws, wood, iron or steel, except coach screws	37.0
185	Woven fabric <85 % artificial staple fibres, printed	35.9
186	Womens, girls overcoats etc., of cotton, not knit	35.1
187	Press-fasteners, snap-fasteners and press-studs, parts	34.8
188	Pneumatic tyre moulding and retreading machinery, etc.	34.0
189	Woven fabric >85 % acrylic staple fibres, nes	32.8
190	Numerically controlled surface grinding machines, accurate to 0.01 mm	32.4
191	Patterns, moulding	32.2
192	Mens, boys overcoats of cotton, not knit	32.0
193	Buttons of base metal, not covered with textile	31.9
194	Parts of frames and mountings for spectacles etc.	31.8
195	Cut or sawn slabs of stone nes	30.8
196	Womens, girls overcoats of material nes, not knit	30.4
197	Chenille, loop whale, gimped (except metallised) yarn	29.9
198	Plain weave cotton, >85 % <100 g/m^2, printed	29.9
199	Synthetic rubber or factice latex from oils, nes	29.4
200	Footwear, soles/uppers leather, strap instep and big toe	29.0
201	Iron oxides and hydroxides	28.6
202	Instruments nes for aeronautical/space navigation	28.5

(continued)

Table 3.11 (continued)

Rank	Commodity description	Trade balance value (million $)
203	Ultra-violet or infra-red ray apparatus	28.5
204	Jacks and hoists except hydraulic and garage hoists	28.0
205	Transfers (decalcomanias), except vitrifiable	27.7
206	Woven fabric >85 % nylon, polyamide, dyed, nes	27.3
207	Keyboard instruments electrical/requiring amplifier	27.2
208	Machinery for sugar refining and manufacture	25.3
209	Plaited bands/etc., iron/steel, no electric insulation	25.2
210	Glass mirrors, framed	24.8
211	Punching, notching, punch-shear machine tools, nes	24.8
212	Thiocarbamates and dithiocarbamates	24.6
213	Textile yarn doubling or twisting machines	24.4
214	Auxiliary machines nes for spin, weave, knit machines	24.1
215	Parts and accessories of shotguns or rifles, nes	23.9
216	Coffee, roasted, decaffeinated	23.0
217	Parts and accessories for flashlights and flashbulbs	22.2
218	Mens, boys overcoats of impregnated, etc., fabric	21.3
219	Metal cuttting, shaping, filing, engrave machines, nes	21.1
220	Spinach fresh or chilled	20.5
221	Numerically controlled shearing (non-punching) machine tools	20.3
222	Flour or meal of oil seed, fruit, except mustard, soya	20.0
223	Shawls, scarves, etc., synthetic fibres, not knit	20.0
224	Articles containing magnesite, dolomite or chromite	19.8
225	Woven fabric synthetic filament, printed, nes	19.8
226	Moulds for glass	19.6
227	Tools for working in the hand, pneumatic type nes	19.3
228	Furniture to take refrigerating/freezing equipment	19.2
229	Lavatory seats and covers of plastics	19.1
230	Tubes of glass other than low expansion	19.1
231	Wire, aluminium alloy, t < 7 mm	18.8
232	Mens, boys overcoats, etc., of material nes, knit	18.6
233	Bobbins, spools and similar supports of paper, nes	17.6
234	Electric capacitors, variable or adjustable (pre-set)	17.6
235	Hydrocarbon derivs, mixed sulpho/nitro/nitroso groups	17.3
236	Pedestrian controlled tractors	17.1
237	Trunks, suit-cases and similar containers, outer nes	16.7
238	Metal broaching machines	16.1
239	Mens, boys overcoats of material nes, not knit	15.6
240	Woven cotton nes, <85 % + manmade fibre, <200 g/m^2 dyed	15.4
241	Hand pumps not designed to measure flow	15.4
242	Aluminium pipe or tube fittings	15.4
243	Lead waste or scrap	15.1

(continued)

Table 3.11 (continued)

Rank	Commodity description	Trade balance value (million $)
244	Trout, live	15.1
245	Babies garments, accessories of material nes, not knit	14.8
246	Woven fabric <85 % artif staple + manmade fil yarn dyed	14.2
247	Woven cotton nes, <85 % + manmade fibre, >200 g/m^2, dyed	14.1
248	Yarn of carded fine animal hair, not retail	13.1
249	Conveyor belts nes	12.7
250	Turkeys, not cut, fresh	12.6
251	Scissors, tailors/similar shears, blades thereof	12.2
252	Worked human hair, wool or animal hair, for wig making	12.2
253	Electric brazing, soldering machines and apparatus nes	12.0
254	Tools, broom handles, bodies, etc., of wood	11.8
255	Cotton yarn (except sewing thread) <85 % cotton, retail	11.7
256	Vitamins A, derivatives, unmixed	11.5
257	Textile fibre drawing or roving machines	10.8
258	Mowers, powered, lawn, nes	10.5
259	Textile wall coverings	10.5
260	Ceramic construction items, chimney pots and cowls	10.5
261	Net/fencing, iron or steel <3 mm wire, <100 cm mesh, nes	10.4
262	Quilted textile products in piece (not embroidered)	10.2
263	Ornamental trimmings in the piece	10.1
264	Washers, iron or steel, except spring/lock	10.1
265	Chain parts, iron or steel, except articulated link	9.9
266	Cream separators	9.3
267	Parts and accessories for the musical instruments nes	9.2
268	Ultrasonic process machine tools	9.2
269	Woven fabric polyester staple fibres, nes	9.1
270	Coriander seeds	8.9
271	Muzzle-loading firearms	8.8
272	Maps and charts, printed, other than book form	8.8
273	Twill cotton except denim, >85 % >200 g/m^2, yarn dyed	8.8
274	Woven fabric synthetic staple fibre with manmade, nes	8.7
275	Metal shaping or slotting machines	8.5
276	Drilling, threading or tapping tools	8.4
277	Red lead and orange lead	8.2
278	Globes, geographical, printed	8.0
279	Flat rolled i/nas, plated or coated with tin, w >600 mm, t >0.5 mm	8.0
280	Flat rolled i/nas, plated or coated width <600 mm	7.9
281	Office supplies nes, of paper, book covers, blotters	7.2
282	Walking-sticks, seat-sticks, whips, etc.	7.1
283	Wirewound variable resistors, rheostats, etc. >20 W	6.8
284	Woven fabric <85 % artificial staple fibre, yarn dyed	6.8

(continued)

Table 3.11 (continued)

Rank	Commodity description	Trade balance value (million $)
285	Tarpaulins, awnings and sunblinds, of material nes	6.6
286	Twill weave cotton, >85 % >200 g/m^2, printed	6.3
287	Womens, girls suits, synthetic fibres, knit	6.2
288	Cuff-links and studs of base metal, plated or not	6.1
289	Salts and esters of tartaric acid	5.2
290	Woven cotton nes, <85 % + manmade fibre, >200 g, bleached	5.2
291	Tubes and pipe, nickel, not alloyed	5.1
292	Other cloth, grill, netting and fencing, of aluminium wire	4.9
293	Statuettes, other ornaments plated with precious metal	4.8
294	Asbestos fabricated products nes	4.8
295	Parts of electronic valve and tubes, except cathode ray	4.6
296	Parts of book-binding and book-sewing machines	4.3
297	Boring-milling machines nes for removing metal	4.3
298	Endless bands for machine	4.2
299	Woven fabric >85 % synth nes + cotton, <170 g/m^2 yarn dye	4.2
300	Unroasted iron pyrites	4.1
301	Womens, girls bathrobe, dressing gowns, nes, knit	4.1
302	Metal and wire cutting shears, etc.	3.9
303	Articles of porcelain or china nes	3.8
304	Yarn of other vegetable textile fibres	3.6
305	Letterpress printing machinery nes except flexographic	2.9
306	Gloves, mittens or mitts, nes, of wool or hair, knit	2.7
307	Typewriters, electric, nes	2.6
308	Yarn of synthetic filament not sewing thread, retail	1.8
309	Cotton yarn >85 % multiple combed >714 dtex, not retail	1.6
310	Wrench sockets, ratchet handles, extensions, spinners	1.4
311	Woven cotton fabric, >200 g/m^2, bleached, nes	1.2
312	Asbestos clothing, accessories, foot and headwear	1.1
313	Ties, bow ties and cravats, manmade fibres, not knit	1.1
314	Skipjack, stripe-belly bonito, fresh or chilled, whole	0.9
315	Woven fabric of coarse animal hair or horse hair	0.8
316	Earth colours with >70 % iron as Ferrous oxide	0.8
317	Calculating machines, non-electric	0.7
318	Woven fabric of jute/bast fibre, not unbleach/bleached	0.7
319	Clocks with watch movements, nes (except vehicle)	0.4
320	Anthraquinone	0.4
321	Cobalt ores and concentrates	0.2

Index of competitive excellence of Italy in the world trade; Fortis-Corradini's Index (case study of a total sample of 5,117 products that comprises International Trade)
Source Compiled by Fondazione Edison on data from Istat, Eurostat and UN Comtrade

Table 3.12 Trade balance: The 1.235 products in which Italy beats Germany; year 2012 (million dollars)

Rank	Commodity description	Italy trade balance (million $)	Germany trade balance (million $)
1	Gold in unwrought forms non-monetary	5,066.9	4,095.5
2	Taps, cocks, valves and similar appliances, nes	4,932.7	3,279.3
3	Jewellery and parts of precious metal except silver	4,281.9	−156.9
4	Motor vehicle parts nes	3,786.9	2,092.2
5	Footwear, outer soles and uppers of leather, nes	2,652.2	−169.8
6	Handbags with outer surface of leather	2,536.0	−227.1
7	Packing or wrapping machinery nes	2,403.2	1,900.7
8	Glazed ceramic flags, tiles wider than 7 cm	2,311.0	−300.1
9	Furniture, wooden, nes	1,877.9	−690.1
10	Furniture parts nes	1,870.7	169.9
11	Articles of iron or steel, nes	1,857.0	1,407.6
12	Sunglasses	1,828.4	−98.7
13	Bovine and equine leather, full or split grain, nes	1,749.0	−37.8
14	Unglazed ceramic flags, tiles > 7 cm wide	1,531.8	1.6
15	Parts of gas turbine engines except turbo-jet/prop	1,438.8	1,048.6
16	Parts of wash, filling, closing, aerating machinery	1,393.2	982.6
17	Helicopters of an unladen weight > 2,000 kg	1,361.6	198.4
18	Electric conductors, 80–1,000 V, no connectors	1,297.6	186.5
19	Seats with wooden frames, upholstered nes	1,217.3	−1,022.0
20	Articles of aluminium, nes	1,211.2	30.1
21	Parts of machines and mechanical appliances nes	1,164.5	785.2
22	Brake system parts except linings for motor vehicles	1,147.0	913.0
23	Pipes and tubing, stainless steel, welded	1,108.7	25.7
24	Hollow profiles/tubes, iron/steel, non-circular, welded	1,081.0	−447.8
25	Flat rolled i/nas, coated with zinc, width >600 mm, nes	1,053.9	−549.3
26	Pumps nes	968.5	−34.4
27	Pipes etc., nes, iron/steel welded nes, diameter <406.4 mm	949.0	−122.2
28	Structures and parts of structures, iron or steel, nes	938.7	753.1
29	Bovine and equine leather, nes	921.6	−10.3
30	Footwear, sole rubber, plastics uppers of leather, nes	918.4	−940.8
31	Bar/rod, i/nas, indented or twisted, nes	862.1	127.0
32	Refrigerator/freezer chests/cabinets/showcases	852.0	−362.9
33	Parts of vacuum pumps, compressors, fans, blowers, hoods	819.4	172.2
34	Other flat, not in coils, not further worked than hot-rolled < 10	777.9	−73.5
35	Commercial equipment, hot drinks/cooking/heating food	747.1	460.7
36	Articles, iron or steel nes, forged/stamped, nfw	721.6	−284.9

(continued)

Table 3.12 (continued)

Rank	Commodity description	Italy trade balance (million $)	Germany trade balance (million $)
37	Bakery and pasta making machinery	707.7	239.3
38	Automatic washing machines, of a dry capacity <10 kg	704.4	−25.2
39	Heat exchange units, non-domestic, non-electric	682.4	483.7
40	Stainless steel bar nfw than cold formed/cold finished	653.0	−184.1
41	Chandeliers, other electric ceiling or wall lights	637.9	36.6
42	Parts of metal rolling mills and rolls	633.8	2.3
43	Sheet/film not cellular/reinf polymers of propylene	626.8	438.7
44	Footwear, soles, uppers of leather, over ankle, nes	618.3	−68.5
45	Pullovers, cardigans etc., of wool or hair, knit	592.0	−309.1
46	Electric cooking, grilling and roasting equipment nes	591.1	503.8
47	Worked marble, travertine and alabaster	583.9	−27.1
48	Woven fabric, >85 % combed wool or fine hair, <300 g/m^2	580.6	18.0
49	Insulated winding wire of copper	569.5	485.2
50	Self-adhesive plastic, rolls <20 cm wide	560.5	146.1
51	Jewellery and parts, silver, including plated silver	559.2	−161.1
52	Lubricating oil additives with petroleum, bitumen oils	534.8	−325.4
53	Articles for pocket or handbag, leather outer surface	525.1	−29.7
54	Engines, spark-ignition reciprocating, over 1,000 cc	520.4	−996.8
55	Cooking appliances for gas fuel, etc., iron or steel	518.8	−58.1
56	Womens, girls ensembles, material nes, not knit	516.6	−1.0
57	Boots, sole rubber or plastic upper leather, nes	515.3	−547.4
58	Machines to agglomerate, shape, mould minerals or fuel	506.2	502.8
59	Machinery for making up pulp, paper, paperboard nes	494.7	136.2
60	Parts of turbo-jet or turbo-propeller engines	481.6	−1,382.2
61	Motorcycles, spark ignition engine of >800 cc	471.3	389.0
62	Worked granite	470.9	−267.9
63	Machine tools for wood, cork or hard plastic, etc., nes	468.7	27.4
64	Womens, girls garments nes, material nes, not knit	466.2	−5.8
65	Pipes, line, iron or steel, for oil or gas pipelines	453.3	323.5
66	Pipe and tube fittings, of copper alloys	448.0	269.5
67	Moulds, injection and compression, for rubber or plastic	438.6	206.3
68	Parts of refrigerating or freezing equipment	429.4	−153.1
69	Wheels including parts/accessories for motor vehicles	429.3	−89.4
70	Fittings, pipe or tube, iron or steel, nes	427.3	293.2
71	Boring or sinking machinery nes, self-propelled	423.1	262.0
72	Paper handkerchiefs, cleansing, facial tissues, towels	422.2	166.9

(continued)

Table 3.12 (continued)

Rank	Commodity description	Italy trade balance (million $)	Germany trade balance (million $)
73	Articles of apparel of leather or composition leather	412.3	−174.0
74	Pipe flanges, iron or steel except stainless/cast	409.6	34.4
75	Engines, diesel, for motor vehicles	404.2	204.1
76	Mens, boys suits, of wool or hair, not knit	402.8	19.8
77	Bedroom furniture, wooden, nes	402.5	144.2
78	Parts of taps, cocks, valves or similar appliances	395.1	318.1
79	Bars, rods and other profiles, aluminium alloyed	391.7	−488.4
80	Electric coffee or tea makers, domestic	391.3	−482.1
81	Motorcycles, spark ignition engine of 50–250 cc	384.2	−119.8
82	Refrigerating or freezing equipment nes	381.8	14.2
83	Panty hose etc., synthetic fibre yarn <67 dtex/sy, knit	377.7	−94.9
84	Ventilating hoods having a maximum width <120 cm	368.1	82.7
85	Belts and bandoliers of leather or composition leather	367.4	−48.4
86	Womens, girls dresses, of material nes, not knit	366.6	−22.6
87	Machine tools nes for stone, ceramics and cold glass	353.7	39.1
88	Parts, accessories nes, metal shaping machine tools	350.9	248.9
89	Parts of lifts, skip hoist or escalators	350.6	−150.2
90	Reservoirs/tanks/vats/etc., iron/steel capacity >300 l	342.5	145.8
91	Cranes designed for mounting on road vehicles	332.3	−39.8
92	Frames and mountings for spectacles etc., of plastic	328.7	−47.0
93	Parts of footwear nes, gaiters and leggings etc.	326.8	−73.5
94	Paper, household, sanitary, width >36 cm	317.3	−85.4
95	Bearings, ball	315.8	−168.3
96	Polyesters nes, unsaturated, in primary forms	315.3	−98.4
97	Cut or sawn slabs of marble, travertine or alabaster	314.5	−16.2
98	Machinery for preparing or making up tobacco	313.8	−125.5
99	Parts of balloons, dirigibles, spacecraft	312.3	−212.7
100	Seats with metal frames, upholstered nes	305.2	−57.0
101	Textured yarn nes, nylon, polyamide >50 dtex not retail	303.7	−8.5
102	Air conditioners nes, without refrigerating unit	303.4	194.6
103	Electric lamps, lighting fittings, nes	292.9	244.2
104	Air conditioners nes, with refrigerating unit	292.7	79.8
105	Handbags with outer surface plastics, textile material	292.6	−269.2
106	Woven fabric, carded wool or hair + manmade staple fiber	288.3	−7.7
107	Table, kitchen and household	287.4	−145.1
108	Propene (propylene)	287.2	−1,092.1
109	Butt weld fittings, iron/steel except stainless/cast	287.1	60.4

(continued)

Table 3.12 (continued)

Rank	Commodity description	Italy trade balance (million $)	Germany trade balance (million $)
110	Patent leather, metallised leather	285.9	−2.4
111	Semi-finished products of alloy steel except stainless	284.1	−80.6
112	Womens, girls overcoats, of impregnated, etc., fabric	280.2	−6.9
113	Soles and heels for footwear, of rubber or plastic	278.5	−20.8
114	Office furniture, wooden, nes	275.4	134.3
115	Washing and cleaning preparations, retail	275.4	258.4
116	Clasps/buckles, etc., for clothing, footwear, bags etc.	270.6	−29.5
117	Furniture of cane, materials nes	269.9	2.9
118	Garments of knit or crochet impregnated fabric	265.8	−36.1
119	Boring-milling machines num controlled for metal	264.2	142.3
120	Domestic iron/steel solid fuel appliances, not cookers	263.6	−96.5
121	Hot rolled stainless steel coil, w >600 mm, t 3–4.75 mm	263.1	−51.4
122	Rubber or plastic moulding and forming machines nes	260.8	101.6
123	Parts for soil preparation or cultivation machinery	260.6	106.5
124	Nonwovens, man-made filaments	258.0	184.8
125	Safety glass (laminated) for vehicles, aircraft, etc.	256.0	−178.1
126	Asphalt or similar material articles, in rolls	252.1	19.0
127	Seats nes	251.8	−110.0
128	Bars and rods, iron, n/f/workd	246.4	−108.4
129	Grinding/polishing machines for stone, ceramics, glass	242.2	90.7
130	Containers for compressed/liquefied gas, iron or steel	237.2	112.2
131	Dish washing machines commercial	235.7	206.6
132	Lighting/visual signalling equipment nes	234.8	−425.7
133	Metal treating machines, electric wire coil-winders	234.8	227.0
134	Watch straps etc., and parts, of/clad with precious metal	233.9	−1.7
135	Toilet paper	232.9	119.1
136	Articles of apparel and clothing accessories of furskin	232.0	20.8
137	Shotguns, shotgun-rifles for sport, hunting or target	230.7	12.9
138	Knit or crochet fabric, width >30 cm, >5 % elastomer	229.8	55.4
139	Door closures, automatic, of base metal	228.3	144.0
140	Eye make-up preparations	226.5	210.1
141	Trays, dishes, plates, cups and the like, of paper	225.3	−41.2
142	Valves, pressure reducing	225.0	136.3
143	Garments nes, of materials nes, knit	222.9	−30.5
144	Garments made up of textile felts and nonwoven fabric	221.4	−54.9
145	Safety glass (tempered) for vehicles, aircraft, etc.	218.0	−240.7
146	Parts of central heating boiler nes	211.4	−204.5

(continued)

Table 3.12 (continued)

Rank	Commodity description	Italy trade balance (million $)	Germany trade balance (million $)
147	Ties, bow ties and cravats, of silk, not knit	210.6	−15.2
148	Parts of seats	210.1	−1,754.4
149	Agricultural sprays and powder dispersers	210.0	66.7
150	Adrenal cortical hormones, in dosage	209.1	−72.0
151	Frames and mountings for spectacles etc., except plastic	208.3	−35.5
152	Rubber tube, pipe, hose, metal reinforced, no fittings	207.8	−9.7
153	Other nonwovens, man-made filaments	207.1	−136.4
154	Supported catalysts, precious metal based	205.8	58.6
155	Earth moving/road making equipment, self-propelled nes	205.7	−45.8
156	Furniture, plastic, nes	204.7	−27.4
157	Wadding, products, manmade fibres, not sanitary items	200.8	32.2
158	Plastic household, toilet articles not table, kitchen	198.2	−39.1
159	Carbon (carbon blacks and other forms of carbon, nes)	198.1	−38.3
160	Sawing machines for stone, ceramics and glass	198.0	−8.0
161	Waste or scrap of unbleached kraft or paperboard	197.9	−48.7
162	Mowers, powered, lawn, with horizontal cutting device	196.4	−325.7
163	Bars and rods, rectangular cross-section	194.0	−320.1
164	Machine tools to forge, stamp, hammer or press metals	192.1	185.1
165	Machine tools to bend, fold, shear or press metal, nes	191.9	119.0
166	Isolating and make-and-break switches, voltage >1 kV	191.4	−44.3
167	Paper, glassine, glazed transparent or translucent nes	191.0	−98.7
168	Drilling or morticing machines for wood, etc.	190.1	64.0
169	Casting machines used in metallurgy, foundries	189.0	16.8
170	Railway/tramway rails, iron or steel	188.5	−51.7
171	Platinum unwrought or in powder form	185.2	−529.9
172	Electric instant, storage and immersion water heaters	183.9	−11.4
173	Automatic beverage-vending machines incorporating heating or refrigerating devices	180.9	0.1
174	Stainless steel sheet, w < 600 mm, cold rolled/reduced	176.3	32.6
175	Plastic sacks, bags, cone except of ethylene polymers	175.7	66.0
176	Sacks and bags, of paper, nes, including cones	174.9	109.4
177	Moulds, injection or compression, for metals/carbides	174.7	−11.0
178	Sports footwear, except ski, uppers of leather	174.4	−45.2
179	Containers nes, outer surface of leather	174.4	−74.7
180	Alkaloids, derivs, no antibiotics, hormones, in dosage	174.3	−227.0

(continued)

Table 3.12 (continued)

Rank	Commodity description	Italy trade balance (million $)	Germany trade balance (million $)
181	Play, fair-ground equipment, travellng circus, theatre	172.1	84.6
182	Rollers, soil preparation, cultivation machinery, nes	172.0	−2.1
183	Parts, accessories nes, stone, ceramic machine tools	167.6	23.5
184	Calendering or rolling machines, not for metals/glass	167.2	88.2
185	Industrial furnace, oven, incinerator non-electric nes	166.4	130.2
186	Office furniture, metal, nes	166.1	37.8
187	Doors, frames and thresholds, of wood	165.0	108.9
188	Machinery to impregnate textiles, make linoleum, etc.	164.5	162.4
189	Womens, girls overcoats etc., of wool or hair not knit	163.2	−52.5
190	Burglar or fire alarms and similar apparatus	163.1	38.8
191	Stranded steel wire/cable/etc., no electric insulation	162.6	63.6
192	Shawls, scarves, etc., of wool or hair, not knit	161.2	−21.8
193	Clothing accessories nes, textile material, not knit	160.6	−10.9
194	Gaskets, washers and other seals of vulcanised rubber	160.4	59.7
195	Non-electric heaters (with fan), parts, of iron/steel	159.5	2.9
196	Prefabricated buildings	158.9	132.6
197	Printing machinery nes	157.8	148.7
198	Scent sprays and similar toilet sprays, parts	156.5	54.6
199	Marble and travertine in blocks etc.	155.7	−3.8
200	Paper, self-adhesive, cut to size, in strips or rolls	155.7	31.9
201	Tower cranes	153.9	138.4
202	Monolithic integrated circuits	153.4	−442.4
203	Distilling or rectifying plant	152.2	104.8
204	Penicillins and streptomycins, derivs, in dosage	151.0	−83.7
205	Parts, accessories nes, wood, plastic machine tools	150.1	136.5
206	Circular knitting machines, diameter <165 mm	149.7	12.7
207	Flat, in coils, not further worked than cold-rolled (cold-reduced) <1>3 m	148.3	−251.7
208	Ingots and other primary forms, stainless steel	148.0	−11.4
209	Nonwovens, man-made filaments >70<25 g/m^2	147.9	−26.2
210	Sacks and bags, of paper, having a width >40 cm	147.7	−23.8
211	Polyvinyl chloride nes, plasticised in primary forms	147.4	99.1
212	Woven fabric, >85 % combed wool or fine hair, >300 g/m^2	146.3	14.9
213	Trunks, suit-cases/etc., outer surface leather	146.2	−11.5
214	Machines for manufacturing or hot working glass or glassware	144.6	48.8
215	Tiles and flagstones of cement, concrete, artificial stone	143.5	47.8

(continued)

Table 3.12 (continued)

Rank	Commodity description	Italy trade balance (million $)	Germany trade balance (million $)
216	Machines to mill or work cereals or dried legumes nes	142.2	43.4
217	Auxiliary plant for steam/vapour generating boilers	141.9	34.7
218	Cans, iron or steel, capacity <50 l nes	140.7	99.7
219	Ceramic bathroom kitchen sanitary items not porcelain	139.1	−5.0
220	Other flat, not in coils, not further worked than hot-rolled <4.7	139.1	−385.2
221	Mens, boys jackets and blazers, wool or hair, not knit	139.0	−49.5
222	Parts for domestic non-electric thermic appliances	137.5	−16.0
223	Marble and travertine, crude or roughly trimmed	137.3	11.9
224	Flanges, stainless steel	137.2	5.6
225	Pipe fittings, butt welding of stainless steel	136.8	68.1
226	Pneumatic tyres new of rubber nes	136.3	−147.5
227	Cans, iron/steel, capacity <50 l closed by crimp/solder	135.4	130.1
228	Vapour generating boilers nes, hybrid boilers	132.8	94.0
229	Scarifiers, cultivators, weeders and hoes	132.7	115.3
230	Machinery to prepare, tan, work hides, skins, leather	131.4	4.4
231	Motorcycles, spark ignition engine of <50 cc	130.0	−80.4
232	Mens, boys ensembles, of material nes, not knit	129.5	−1.1
233	Bicycle parts nes	129.4	−288.3
234	Children's picture, drawing or colouring books	129.0	−20.7
235	Animal or vegetable fertilizers, in packs >10 kg	127.2	9.3
236	Womens, girls dresses, of artificial fibres, not knit	126.9	−36.7
237	Machines for balancing mechanical parts, nes	125.2	117.8
238	Corks, crown, of base metal	124.4	24.7
239	Vaccines, human use	124.2	−5.6
240	Watch cases of, or clad with, precious metal	122.8	−23.5
241	Parts of steam or vapour generating boilers nes	121.5	−3.1
242	Generating sets, diesel, output <75 kVA	121.4	30.7
243	Moulds for rubber or plastic, nes	120.2	48.4
244	Machinery for preparation of fruits, nuts, vegetables	120.2	11.7
245	Physical exercise, gymnasium and athletics equipment	119.8	−148.0
246	Mountings, fittings, of base metal, nes	118.9	112.4
247	Built-in jacking systems for garages	118.9	78.8
248	Parts of equipment for metallurgy, foundries	118.5	118.3
249	Shearing (except punch-shear) machine tools, nes	117.6	60.5
250	Parts of fork-lift etc., trucks	117.4	−26.7
251	Plastic articles for goods conveyance or packing nes	116.8	57.5
252	Plastic table and kitchen ware	116.6	−162.7

(continued)

Table 3.12 (continued)

Rank	Commodity description	Italy trade balance (million $)	Germany trade balance (million $)
253	Ski-boots, cross country, ski footwear and snowboard boots	116.5	−29.4
254	Compounded unvulcanised rubber in plate, sheet, strip	115.8	72.1
255	Bar/rod, cold formed/finished free cutting steel	115.7	−27.0
256	Woven cotton nes, >85 % <200 g/m^2, yarn dyed	115.3	2.5
257	Pullovers, cardigans etc., of material nes knit	115.1	−37.7
258	Waste or scrap containing gold as sole precious metal	113.9	−1,318.1
259	Bakery ovens, etc., non-electric	113.7	36.1
260	Registers, account books, note books, diaries etc.	113.7	−10.5
261	Washing, bleaching or dyeing machines (non-domestic)	113.0	92.2
262	Tube, pipe or hose, rigid, of polyvinyl chloride	112.6	57.2
263	Medical, surgical or laboratory sterilizers	111.9	74.2
264	Womens, girls garments nes, of impregnated fabric	111.5	−119.0
265	Wire, iron non-alloy steel, not plated or coated, not polished	110.7	57.5
266	Tartaric acid	110.6	−12.2
267	Knit or crochet fabric of manmade fibres, nes	110.1	45.0
268	Parts of watch cases	109.1	−19.4
269	AC generators, of an output <75 kVA	108.8	−28.7
270	Shawls, scarves, etc., of silk etc., not knit	108.6	−27.8
271	Footwear, uppers of leather with metal toe-cap, nes	108.4	−67.4
272	Pipe fittings of non-malleable cast iron	108.1	17.4
273	Womens, girls dresses, of cotton, not knit	107.3	−86.4
274	Woven fabric >85 % artificial filament/strip, dyed, nes	106.0	12.5
275	Electric table, desk, bedside and floor lamps	105.5	−128.3
276	Womens, girls trousers and shorts, of cotton, not knit	104.7	−932.6
277	Yarn of carded wool, >85 % wool, not retail	104.2	6.7
278	Palladium unwrought or in powder form	103.9	−145.7
279	Yarn of carded, wool, <85 % wool, not retail	103.8	−4.1
280	Powders, for skin care and make-up	102.9	−13.0
281	Bars, rods and profiles, copper alloy nes	101.8	19.8
282	Precious metal compounds except gold, silver, amalgams	101.7	72.7
283	Footwear, sole leather and uppers textile material	101.2	−10.9
284	Paper, fine, woodfree, >150 g/m^2, uncoated, nes	100.6	8.4
285	Wire, plated or coated with other base metal	100.6	21.1
286	Ethylene	100.3	−494.2
287	Non-automatic electric plasma and other arc welders	99.7	27.3
288	Bar/rod, iron or non-alloy steel, nes	99.1	−31.9
289	Circular saw blades, working part of steel	97.8	28.3

(continued)

Table 3.12 (continued)

Rank	Commodity description	Italy trade balance (million $)	Germany trade balance (million $)
290	Denim cotton >85 % >200 g/m^2	97.5	0.4
291	Glues or adhesives, prepared nes, package >1 kg	95.3	12.0
292	Articles of glass, nes	94.5	42.5
293	Machinery for producing or preparing textile yarn nes	94.5	11.2
294	Dies for drawing or extruding metal	94.1	14.5
295	Bicycle saddles	93.8	−29.6
296	Parts of household or laundry-type washing machines	93.4	64.7
297	Woven fabric, >85 % carded wool or fine hair, >300 g/m^2	93.3	−1.4
298	Womens, girls suits, of material nes, not knit	93.2	−2.3
299	Furnace burners for liquid fuel	91.7	88.6
300	Other bars and rods, not further worked than cold-formed or cold-finished	91.5	−103.0
301	Parts of electric accumulators, including separators	91.4	−43.7
302	Wire of alloy steel nes	91.3	−113.2
303	Parts and accessories for photographic cameras	90.8	3.4
304	Moulds for metal or metal carbides, nes	90.7	−0.9
305	Grill netting and fencing, welded, i/nas <3 mm wire, <100 cm mesh	90.6	88.7
306	Panty hose etc., synthetic fibre yarn >67 dtex/sy, knit	90.4	−19.8
307	Flat rolled products of iron or non-alloy steel <1>	90.1	−66.5
308	Tube mills, metal rolling	89.6	31.6
309	Soaps nes	89.5	31.9
310	Teleferiques, chair-lifts, ski-lifts, etc.	89.4	8.1
311	Womens, girls blouses and shirts, of silk, not knit	89.2	−15.1
312	Flat rolled products of iron or non-alloy steel <4>3	89.2	−100.0
313	Machines to process animal or fixed veg fats or oils	88.8	69.9
314	Grindstones etc., of synthetic or natural diamond	88.7	4.6
315	Table/kitchen articles, parts, of iron or steel, nes	88.4	−18.9
316	Fluoro-polymers nes in primary forms	88.0	−56.3
317	Motorcycle parts except saddles	87.1	−267.7
318	Rubber tube, pipe, hose textile-reinforced no fittings	86.3	74.6
319	Parts etc., of spinning, combing, drawing machines nes	86.2	74.9
320	Seats with metal frames, nes	85.7	−261.4
321	Womens, girls dresses, synthetic fibres, not knit	85.4	−132.7
322	Machinery for making or repairing footwear	84.5	16.7
323	Woven cotton nes, >85 % >200 g/m^2, dyed, nes	84.2	36.1

(continued)

Table 3.12 (continued)

Rank	Commodity description	Italy trade balance (million $)	Germany trade balance (million $)
324	Finishing agents and dye carriers—textile industry	84.1	66.8
325	Tube/pipe/hollow profile, iron/steel, riveted/open seam	83.6	14.9
326	Automatic goods vending machines without heating/refriger. Devices; money changing machines	83.3	17.5
327	Handbags, of vulcanised fibre or paperboard	83.0	−3.9
328	Electric signal, safety and traffic controller parts	82.8	51.6
329	Paper, kraft, >95 % chem pulp, <150 g, bleached, clay coat	82.6	−55.0
330	Loudspeakers, nes	82.3	−150.1
331	Parts and accessories equally suitable for use with machines	82.2	−1.8
332	Parts of machinery for leather industry	81.5	14.9
333	Mens, boys trousers and shorts, wool or hair, not knit	81.4	−14.4
334	Pipe fittings of malleable iron or steel, cast	81.4	−25.4
335	Woven fabric, combed wool or hair with natural fibres	81.0	1.8
336	Bars, rods and profiles, nickel alloy	80.8	63.5
337	Sound signalling equipment	80.6	10.1
338	Winches or capstans nes powered by electric motor	80.6	70.8
339	Ironing machines and presses including fusing presses	80.3	64.7
340	Mens, boys garments nes, made up of impregnated fabric	80.1	−143.7
341	Semi-finished product, iron/nas >0.25 %C	79.9	78.7
342	Tankers	79.6	−166.0
343	Flat rolled i/nas, painted/plastic coated, width >600 mm	79.5	−225.6
344	Machinery to reel, fold, cut, pink, etc., textile fabric	79.0	40.5
345	Presses, crushers etc., for wine, fruit juice, beverages	78.9	17.2
346	Rowing boats, canoes, pleasure boats except sail/power	78.7	3.2
347	Dump trucks designed for off-highway use	78.0	72.5
348	Winches or capstans nes	77.8	56.2
349	Radiators and parts thereof, iron or steel except cast	77.7	24.7
350	Threshing machinery nes	77.1	0.4
351	Worked calcareous stone nes	77.1	43.4
352	Mattresses of cellular rubber or plastic	77.0	−119.3
353	Oral and dental hygiene preparations, except dentifrices	76.5	63.5
354	Safety headgear	76.2	−33.4
355	Cartridges, shotgun	76.1	−6.6
356	Bars, rods and profiles, aluminium, not alloyed	75.6	−43.0

(continued)

Table 3.12 (continued)

Rank	Commodity description	Italy trade balance (million $)	Germany trade balance (million $)
357	Pipe-line longitudinal nes welded steel, diameter >406 mm	75.5	40.2
358	Parts, accessories for gas, liquid, electricity meters	74.6	8.0
359	Wallpaper and similar wall coverings, nes	74.1	60.5
360	Womens, girls jackets, blazers, wool or hair, not knit	74.1	−13.9
361	Table/kitchen articles, parts, stainless steel	73.8	−69.5
362	Mens, boys overcoats of wool or hair, not knit	73.5	−27.5
363	Shawls, scarves, etc., of material nes, not knit	72.9	−51.9
364	Compounded unvulcanised rubber in primary forms nes	72.9	43.3
365	Heterocyclic compounds with unfused pyridine ring, nes	72.6	−542.7
366	Seats with wooden frames, nes	72.4	−176.4
367	Drilling machines for metal, except num controlled	72.4	11.4
368	Woven fabric of silk, nes	72.4	1.9
369	Egg albumin, dried	71.4	−0.9
370	Woven fabric, >85 % carded wool or fine hair, <300 g/m^2	70.4	−1.3
371	Womens, girls dresses, of artificial fibres, knit	69.7	−39.2
372	Chamois (including combination chamois) leather	69.3	−1.9
373	Worked monumental or building stone nes	69.1	−16.5
374	Hosiery nes, synthetic fibres, knit	68.6	−31.6
375	Electrical fuses, for <1,000 V	68.5	20.6
376	Soaps for purposes other than toilet soap, solid	68.2	7.5
377	Fixed wing aircraft, unladen weight 2,000–15,000 kg	67.6	−151.0
378	Glass cubes, mosaic tiles	67.4	−5.3
379	Flat, electrolytically plated or coated with zinc	67.3	−475.6
380	Clothing accessories nes, of leather or composition	67.2	−0.2
381	Water-skis, surf-boards, other watersport equipment	66.6	−24.1
382	Baby carriages and parts thereof	66.6	−51.5
383	Articles of iron or steel wire, nes	66.5	−31.1
384	Hot rolled, plated or coated with aluminium , w <600 mm, t<3 mm myp275/t>3 mm myp355	66.5	7.3
385	Woven twill >85 % polyester + cotton, >170 g/m^2 dyed	66.4	22.1
386	Woven fabric, >85 % flax, except unbleached or bleached	66.2	1.7
387	Perfumery, cosmetic or toilet preparations, nes	65.4	−1.7
388	Sections, L/T, i/nas, nfw hot-roll/drawn/extruded >80 mm	65.3	−77.9
389	Gravure printing machinery	65.3	48.5

(continued)

Table 3.12 (continued)

Rank	Commodity description	Italy trade balance (million $)	Germany trade balance (million $)
390	Universal AC/DC motors of an output <37.5 W	64.8	−8.9
391	Mens, boys suits, of material nes, not knit	64.8	−1.3
392	Cyclan-, cyclen-, cycloterpen-poly-carboxylic acids	64.8	−50.2
393	Woven fabric >85 % textured polyester, printed, nes	64.4	25.1
394	Parts of garments, clothing accessories nes, not knit	63.8	12.6
395	Brushes, parts, nes	63.6	20.9
396	Synthetic organic tanning substances	62.9	36.4
397	Sinks and wash basins, stainless steel	62.9	54.8
398	Other automatic vending machines incorporating heating/ refrigerating devices	62.6	10.1
399	Flat rld prod/coils >3 mm	62.3	−343.5
400	Womens, girls trousers, shorts, material nes, not knit	62.2	−46.7
401	Flat rolled i/nas, <600 mm, coated with zinc, nes	61.9	−53.8
402	Dry-cleaning machines	61.5	5.0
403	Womens, girls dresses, of wool or hair, not knit	61.3	−3.5
404	Knit, crochet fabric of wool or fine animal hair, nes	61.0	4.5
405	Sections, L, i/nas, nfw hot-roll/drawn/extruded <80 mm	60.6	−48.7
406	Plain weave cotton, >85 % 100–200 g/m^2, yarn dyed	60.4	3.8
407	Balls, needles and rollers for bearings	60.2	4.2
408	Womens full, kneelength hosiery, yarn <67 dtex/sy, knit	60.0	−29.8
409	Womens, girls ensembles, of cotton, not knit	60.0	−4.5
410	Keys, including blanks for keys, of base metal	59.9	43.4
411	Angles, shapes and sections, not further worked than cold-formed or cold-finished	59.9	44.5
412	Threaded fittings, iron or steel except stainless/cast	59.9	28.8
413	Marble granules, chipping and powder	59.5	−216.9
414	Cold metal rolling mills	59.0	33.7
415	Bridges and bridge sections, iron or steel	58.5	45.5
416	Grinding, sanding, polishing machines for wood, etc.	58.5	55.6
417	Parts of automatic goods-vending machinery	58.4	13.6
418	Electrical insulators of glass	58.4	−1.0
419	Unrefined copper, copper anodes, electrolytic refining	58.0	−373.8
420	Dryers for agricultural products	57.8	52.5
421	Bars and rods of iron or non-alloy steel, only hot rolled	57.6	−221.3
422	Grinding machines nes, accurate to 0.01 mm	57.4	6.5
423	Woven fabric of flax, <85 % flax, except unbl/bleached	57.2	2.7
424	Aluminium sanitary ware, parts thereof	57.1	9.4

(continued)

Table 3.12 (continued)

Rank	Commodity description	Italy trade balance (million $)	Germany trade balance (million $)
425	Electrodes etc., of base metal or metal carbide, nes	57.0	2.1
426	Dairy machinery	56.9	55.1
427	Air cond used in vehicle	56.7	55.8
428	Towers and lattice masts, iron or steel	56.4	−236.9
429	Freezers of the chest type, <800 l capacity	56.3	−27.9
430	Flat, not further worked than hot-rolled, plated or coated with aluminium <4.75	56.1	39.1
431	Household, hospital and sanitary supplies, of paper, nes	55.9	−8.4
432	Womens, girls jackets and blazers, material nes, not knit	55.5	−28.7
433	Sulphonated hydrocarbons, salts and ethyl esters	55.3	6.6
434	Womens, girls jackets and blazers, of cotton, not knit	55.1	−106.3
435	Cut or sawn slabs of granite	55.1	−22.3
436	Equipment nes for sports, swimming and paddling pools	54.9	−56.9
437	Plate, sheet, strip of vulcanised cellular rubber	54.7	21.7
438	Flat, in coils, not further worked than cold-rolled (cold-reduced) <3 mm	54.4	−13.8
439	Textile fibre preparing machines nes	54.4	−19.2
440	Pneumatic power engines/motors, except linear acting	54.0	37.0
441	Woven fabric of artificial filament, dyed, nes	54.0	2.0
442	Yarn of wool or fine hair, <85 % such fibres, retail	54.0	−30.1
443	Transformers electric, power capacity 16–500 KVA	54.0	−3.0
444	Pumps fitted, designed to have, a measuring device nes	53.9	18.0
445	Fire extinguishers, whether or not charged	53.7	52.5
446	Vitrifiable enamels, glazes etc., for ceramics and glass	53.6	13.9
447	Bars and rods, not further worked than hot-rolled, hot-drawn or extruded	53.2	−48.8
448	Abrasive powder or grain on a base of other material	52.6	24.3
449	Palmitic acid, stearic acid, their salts and esters	52.4	−14.8
450	Woven cotton nes, <85 % + manmade fibre, >200 g, yarn dyed	52.4	−0.1
451	Waste or scrap of paper or board of mechanical pulp	52.2	−141.0
452	Fixed power capacitors (50/60 Hz circuits)	52.2	19.8
453	Machines making paper/board boxes etc., except moulded	52.1	8.8
454	Liquid dielectric transformers 650–10,000 KVA	50.9	−97.7
455	Womens, girls blouses and shirts, of cotton, not knit	50.4	−349.6
456	Hot rolled stainless steel coil, w > 600 mm, t < 3 mm	49.4	−3.5
457	Label, badge, etc., of woven textile not embroidered	49.1	17.0

(continued)

Table 3.12 (continued)

Rank	Commodity description	Italy trade balance (million $)	Germany trade balance (million $)
458	Air conditioners nes with reverse cycle refrigeration	48.8	−47.0
459	Flat, not further worked than hot-rolled, plated or coated with aluminium, 4 faces<150	48.3	−59.7
460	Sheep or lamb skin leather, nes	48.3	−4.4
461	Optical fibres, except for telecommunications	48.1	−0.5
462	Generating sets, diesel, output 75–375 kVA	47.3	46.8
463	Twill weave cotton, >85 % <200 g/m^2, dyed	47.2	9.0
464	Carboxylic acids (alcohol function only), derivatives	47.2	−41.8
465	Mens, boys garments nes, of material nes, not knit	47.2	−2.6
466	Dentists, barbers or similar chairs and parts	47.1	−2.3
467	Mens, boys shirts, of cotton, not knit	46.9	−489.9
468	Buttons, nes	46.2	7.6
469	Corrugated sheets of asbestos, cellulose fibre cement	46.2	−11.2
470	Electric hair dryers	45.8	−32.6
471	Waterproof footwear (Wellington) no toe cap, over ankle	45.7	−42.6
472	Grill, netting, fencing, ir	45.7	−14.2
473	Womens, girls trousers, shorts, wool or hair, not knit	45.7	−4.4
474	Parts agricultural, forestry, bee-keeping machines nes	45.4	10.3
475	Fixed wing aircraft, unladen weight <2,000 kg	45.4	28.7
476	Camel-back strips for retreading rubber tyres	44.6	1.9
477	"Long pile" knit or crochet textile fabric	44.5	5.4
478	Expanded metal, i/nas <3 mm wire, <100 cm mesh	44.4	1.8
479	Asphalt or similar material articles not in rolls	44.1	1.7
480	Knit or crochet fabric, width < 30 cm, > 5 % elastomer	43.9	7.3
481	Refractory bricks etc., nes	43.8	17.7
482	Footwear, nes	43.4	−39.5
483	Paper, sulphite wrapping, uncoated	43.1	15.2
484	Angles, shapes and sections, of iron or non-alloy steel	42.8	1.0
485	Womens, girls skirts, of material nes, not knit	42.1	−11.0
486	Mats of glass fibres	41.6	31.7
487	Machines for working metal wire	41.4	−16.2
488	Pocket-watch, base-metal case, non-battery	41.2	1.6
489	Woven fabric, elasticised, <30 cm wide	41.0	24.2
490	Glass table or kitchenware, except low expansion glass	40.5	−14.9
491	Milling machines nes, for metal	40.5	27.0
492	Gelatin and derivatives, isinglass, glues (animal) nes	40.2	30.2
493	Angles, shapes and sections, stainless steel	39.8	29.4
494	Hard rubber (e.g. ebonite) in all forms, articles, scrap	39.7	−2.1

(continued)

Table 3.12 (continued)

Rank	Commodity description	Italy trade balance (million $)	Germany trade balance (million $)
495	Water tube boilers, steam production >45 T per hour	39.4	6.0
496	Bar/rod of silico-manganese steel not in coils	39.4	15.4
497	Essential oils of citrus fruits, nes	39.3	−1.7
498	Panty hose etc., of materials nes, knit	39.2	−45.0
499	Threading or tapping machines nes for metal	39.1	15.3
500	Mens, boys jackets and blazers, material nes, not knit	38.9	−12.8
501	Parts and accessories of revolvers or pistols	38.9	11.2
502	Shawls, scarves, etc., of artificial fibres, not knit	38.7	−32.7
503	Tailors dummies, display dummies and automata	38.4	−3.4
504	Bulldozers and angledozers, crawler type	38.2	−34.7
505	Other instruments and apparatus for measuring or checking semiconductor wafers or devices	38.1	−23.7
506	Copper screws, bolts or nuts except wood screws	37.9	15.6
507	Rubberised woven textile fabric, except adhesive tape	37.8	30.4
508	Nitrogen-phosphorus-potassium fertilizers, pack >10 kg	37.6	−80.2
509	Parts grain, seed, dry legumes processing equipment	37.6	16.1
510	Womens, girls overcoats etc., manmade fibre, not knit	37.5	−156.7
511	Mattresses, stuffed, spring interior, etc.	37.4	−51.4
512	Mens, boys ensembles, of cotton, not knit	37.4	−0.7
513	Threaded elbows, bends and sleeves of stainless steel	37.1	−6.8
514	Parts of electric filament or discharge lamps	37.1	6.9
515	Semi-finished product, iron/nas <0.25 %C, nes	36.8	−171.0
516	Sulphonamides in bulk	36.6	−58.4
517	Twill weave cotton, <85 % + manmade fibre, >200 g, dyed	36.5	13.3
518	Gold, silversmith wares, base clad with precious metal	36.0	0.4
519	Woven fabric <85 % artificial staple fibres, printed	35.9	4.9
520	Sheet/film not cellular/reinf cellulose acetate	35.7	30.0
521	Hot rolled stainless steel flat, w > 600 mm, t 3–4.75 mm	35.7	0.5
522	AC generators, of an output 75–375 kVA	35.4	−8.8
523	Flat, not in coils, not further worked than cold-rolled (cold-reduced) <.5 mm	35.1	−29.0
524	Womens, girls overcoats etc., of cotton, not knit	35.1	−63.4
525	Sheet/film not cellular/reinf polyesters nes	35.0	22.3
526	Trucks with lifting/handling equipment, non-powered	34.9	−24.0
527	Silver wares, silver ware plated with precious metal	34.6	2.3
528	Reel fed letterpress printers except flexographic	34.4	0.8
529	Insulated winding wire, nes	34.3	4.6
530	Saw blades nes, including stone cutting, etc.	34.0	17.3

(continued)

Table 3.12 (continued)

Rank	Commodity description	Italy trade balance (million $)	Germany trade balance (million $)
531	Orthophthalic acid esters, nes	33.7	−26.9
532	Woven polyester fab >=85 %	33.5	12.4
533	Articles of leather and composition leather, nes	33.3	0.3
534	Womens, girls dresses, of wool or hair, knit	33.0	−5.0
535	Articles of copper, cast/moulded/stamped, nfw	33.0	−23.4
536	Woven fabric >85 % acrylic staple fibres, nes	32.8	−23.6
537	Woven fabric >85 % silk (except noil silk)	32.5	2.5
538	Parts/access, machines for preparing manmade textiles	32.4	28.9
539	Patterns, moulding	32.2	3.4
540	Paper, clay coated, nes	32.1	30.0
541	Mens, boys overcoats of cotton, not knit	32.0	−29.7
542	Woven cotton nes, >85 % <200 g/m^2, printed	31.9	19.3
543	Buttons of base metal, not covered with textile	31.9	10.6
544	Tyre cord fabric of viscose rayon	31.8	−28.5
545	Parts of frames and mountings for spectacles etc.	31.8	8.1
546	Womens, girls dresses, of material nes, knit	31.7	−7.6
547	Baths, shower-baths and wash basins, of plastics	31.3	−26.5
548	Filtering/purifying machinery/apparatus for beverages	31.2	10.2
549	Essential oils of lemon	31.2	1.8
550	Nitrile-function compounds, nes	31.1	−427.7
551	Sodium sulphites	30.8	−2.9
552	Cut or sawn slabs of stone nes	30.8	−1.8
553	Shawls, scarves, veils etc., textile material, knit	30.7	−100.4
554	Yarn of wool or fine hair, >85 % such fibres, retail	30.5	−13.6
555	Womens, girls overcoats of material nes, not knit	30.4	−5.3
556	Yarn >85 % nylon, etc., staple fibre, multiple, not retail	30.1	−7.1
557	Chenille, loop whale, gimped (except metallised) yarn	29.9	0.9
558	Plain weave cotton, >85 % <100 g/m^2, printed	29.9	11.7
559	Locks of a kind used for motor vehicles of base metal	29.7	−101.4
560	Safes/strong-boxes etc., of base metal	29.4	−50.1
561	Felt, impregnated, coated, covered, or laminated	29.0	1.3
562	Copper-tin base alloys, unwrought	29.0	−25.2
563	Footwear, soles/uppers leather, strap instep and big toe	29.0	−6.1
564	Instantaneous/storage water heaters, not electric nes	28.8	−24.3
565	Natural steatite, crushed or powdered	28.7	−106.1
566	Liquid elevators	28.6	−1.3

(continued)

Table 3.12 (continued)

Rank	Commodity description	Italy trade balance (million $)	Germany trade balance (million $)
567	Iron oxides and hydroxides	28.6	−38.7
568	Ultra-violet or infra-red ray apparatus	28.5	25.7
569	Office binder/file fittings, of base metal	28.4	−39.4
570	Statuettes and other ornaments, base metal, unplated	28.4	−70.7
571	Yarn of acrylic staple fibre and wool or hair, not retail	28.3	−3.4
572	Womens, girls suits, of cotton, not knit	28.2	−8.2
573	Way-type unit head machines, metal working	28.1	0.5
574	Footwear, nes, uppers leather	28.1	−142.9
575	Rubber hygienic, pharmacy items except contraceptives	27.9	25.5
576	Hot rolled stainless steel flat, w > 600 mm, t 4.75–10 mm	27.7	−32.6
577	Stainless steel pipe or tubing, cold rolled	27.7	−36.2
578	Mens, boys anoraks etc., of wool or hair, not knit	27.7	−5.8
579	Yarn, nylon/polyamide, single untwisted nes, not retail	27.5	12.4
580	Other flat, not in coils, not further worked than hot-rolled >3 mm	27.5	−19.1
581	Yarn of combed fine animal hair, not retail	27.5	−2.9
582	T-shirts, singlets etc., of material nes, knit	27.5	−526.6
583	Articles of furskin except clothing and accessories	27.4	−11.3
584	Woven fabric >85 % nylon, polyamide, dyed, nes	27.3	5.4
585	Keyboard instruments electrical/requiring amplifier	27.2	−72.9
586	Copper alloys, unwrought (other than master alloys)	27.2	−67.0
587	Mens, boys jackets and blazers, material nes, knit	27.1	−0.3
588	Woven fabric, combed wool or hair + manmade staple fib	27.1	2.6
589	Womens, girls overcoats, etc., of wool or hair, knit	26.9	−5.6
590	Inflatable pleasure craft	26.9	−13.1
591	U, I or H sections, not further worked than hot-rolled, hot-drawn or extruded <80 mm	26.8	−15.8
592	Polymers, natural or modified nes in primary forms nes	26.8	−63.5
593	Ash or residues containing mainly copper	26.3	−66.6
594	Sheep or lamb skins, raw, wool on, except Persian etc.	26.2	5.5
595	Woven fabric >85 % textured polyester, yarn dyed, nes	26.2	5.9
596	Stone setts, curbstones, flagstones (except slate)	26.1	−84.3
597	Woven fabric <85 % artificial staples, yarn dyed, nes	26.0	0.0
598	Stone mosaic tiles, artificial coloured chips etc.	26.0	−7.6
599	Woven fabric synthetic filament, yarn dyed, nes	26.0	4.7

(continued)

Table 3.12 (continued)

Rank	Commodity description	Italy trade balance (million $)	Germany trade balance (million $)
600	Welding machinery not gas-operated	25.8	3.3
601	Room perfuming or deodorizing preparations, nes	25.7	−15.1
602	Ingot moulds and ladles used in metallurgy/foundries	25.6	25.2
603	Copper-zinc base alloys, unwrought	25.4	−29.0
604	Woven fabric >85 % polyester staple fibres, nes	25.4	14.6
605	Wire, plated or coated with zinc	25.3	−60.6
606	Woven cotton fabric, >200 g/m^2, dyed, nes	25.3	−0.6
607	Machinery for sugar refining and manufacture	25.3	20.5
608	Tulle, lace, embroidery, trimmings etc., making machines	25.3	18.1
609	Machinery for leather, skin goods making except sewing	25.2	0.6
610	Plaited bands/etc., iron/steel, no electric insulation	25.2	−15.5
611	Woven fabric >85 % synthetic filament, dyed, nes	25.2	−0.1
612	Engines, spark-ignition reciprocating, 50–250 cc	25.0	1.7
613	Womens, girls skirts, of cotton, not knit	25.0	−66.0
614	Imitation jewellery nes	25.0	−89.3
615	Wrist-watch, precious metal, battery, other	24.9	0.2
616	L or T sections, not further worked than hot-rolled, hot-drawn or extruded <80 mm	24.9	−4.2
617	Glass mirrors, framed	24.8	−30.3
618	Organic surface-active agents, nes	24.8	19.5
619	Oleic, linoleic or linolenic acids, salts and esters	24.7	−17.1
620	Womens, girls dresses, of synthetic fibres, knit	24.7	−88.3
621	Oestrogens and progestogens, in bulk	24.6	−651.1
622	Thiocarbamates and dithiocarbamates	24.6	1.0
623	Electrical resistors, fixed, except heating, >20 W	24.4	24.0
624	Yarn nylon/polyamide filament multiple nes, not retail	24.2	−9.9
625	Auxiliary machines nes for spin, weave, knit machines	24.1	8.8
626	Jewellery, parts, base metal clad with precious metal	24.1	0.3
627	Spectacles, goggles except sunglasses	24.1	2.2
628	Womens, girls suits, of wool or hair, not knit	24.0	−1.2
629	Textile yarn/strip, rubber, plastic impregnated/coated	24.0	13.7
630	Knit or crochet fabric of other materials, nes	24.0	2.6
631	Sheet/film not cellular/reinf unsaturated polyesters	23.6	−0.5
632	Stearic acid	23.5	−56.3
633	Styrene	23.2	−532.1
634	Parts of slide fasteners	23.1	−1.2
635	Plain weave cotton, >85 % <100 g/m^2, yarn dyed	22.9	0.4

(continued)

Table 3.12 (continued)

Rank	Commodity description	Italy trade balance (million $)	Germany trade balance (million $)
636	Articles of cellular rubber	22.8	9.7
637	Machines for moulding articles in pulp, paper, board	22.5	7.5
638	Waste/scrap containing platinum as sole precious metal	22.3	−968.9
639	Parts and accessories for flashlights and flashbulbs	22.2	−0.2
640	Chain, iron or steel, nes	21.9	1.2
641	Transmission belts or belting, endless synchronous belts	21.6	2.6
642	Flat, in coils, not further worked than hot-rolled, with patterns in relief	21.3	0.8
643	Carpets of felt of textile materials, >0.3 m^2	21.3	−73.9
644	Mens, boys overcoats of impregnated, etc., fabric	21.3	−3.4
645	Castors of base metal	21.3	18.0
646	Metal cuttting, shaping, filing, engrave machines, nes	21.1	9.7
647	Parts for spark-ignition aircraft engines	21.0	0.7
648	Woven fabric synthetic filament, dyed, nes	20.8	4.7
649	Glassware, not kitchen or table ware, not lead crystal	20.8	−80.4
650	Womens, girls skirts, of wool or hair, not knit	20.8	−3.3
651	Linoleum	20.8	12.7
652	Ion exchangers based on polymers in primary forms	20.7	−80.9
653	Phthalic anhydride	20.6	−57.8
654	Yarn of combed wool, <85 % wool, not retail	20.5	−6.2
655	Womens, girls blouses, shirts, manmade fibre, not knit	20.4	−215.4
656	Cinematograph film, exposed and developed, width >35 mm	20.4	−0.5
657	Mens, boys shirts, of materials nes, knit	20.3	−2.7
658	Mechanical lace, manmade fibre (piece, strip, motif)	20.2	2.0
659	Shawls, scarves, etc., synthetic fibres, not knit	20.0	−68.1
660	Powders, copper, of non-lamellar structure	20.0	−6.5
661	Garments nes, of cotton, knit	20.0	−84.3
662	Woven fabric of artificial filament, printed, nes	19.9	1.3
663	Articles containing magnesite, dolomite or chromite	19.8	−21.9
664	Woven cotton fabric, >200 g/m^2, yarns mixed colours	19.8	0.4
665	Woven fabric polyester + manmade filament, nes	19.8	9.7
666	Woven fabric synthetic filament, printed, nes	19.8	−3.1
667	Bovine hides, raw, nes	19.7	11.0
668	Moulds for glass	19.6	−6.3
669	Cooking appliances for solid fuel, iron or steel	19.5	−53.0
670	Woven fabric acrylics + wool or hair, nes	19.4	−0.7
671	Tools for working in the hand, pneumatic type nes	19.3	17.2
672	Furniture to take refrigerating/freezing equipment	19.2	0.4

(continued)

Table 3.12 (continued)

Rank	Commodity description	Italy trade balance (million $)	Germany trade balance (million $)
673	Womens, girls ensembles, of wool or hair, not knit	19.2	−0.3
674	Lavatory seats and covers of plastics	19.1	−1.9
675	Tubes of glass other than low expansion	19.1	−18.9
676	Womens, girls panties, bathrobes, etc., nes not knit	19.0	−1.1
677	Wire, aluminium alloy, t < 7 mm	18.8	−2.3
678	Bicycles, other cycles, not motorized	18.8	−170.0
679	Hosiery nes, of materials nes, knit	18.7	−7.0
680	Leather, of animals nes	18.7	4.4
681	Womens, girls jackets and blazers, of wool or hair, knit	18.7	−11.5
682	Mens, boys overcoats, etc., of material nes, knit	18.6	−3.1
683	Woven fabric <85 % artificial staples, dyed, nes	18.4	4.5
684	Ties, bow ties and cravats, material nes, not knit	18.4	0.1
685	Hosiery nes, of wool or fine animal hair, knit	18.4	−2.4
686	Parts of machinery for preparing tobacco	18.3	−22.6
687	Parts nes for rolling machines, except metals, glass	18.3	7.5
688	Step and repeat aligners	18.3	−221.8
689	Tubes and pipe, aluminium, not alloyed	18.2	−1.2
690	Combined refrigerator-freezers, two door	18.1	−124.4
691	Net/fencing, plastic-coated i/nas <3 mm wire, <100 cm mesh	18.1	3.9
692	Machines to assemble electric lamps, glass envelope	18.1	5.4
693	Hooks, eyes, eyelets, for clothing, footwear, bags etc.	17.9	7.3
694	Converters used in metallurgy or metal foundries	17.9	5.6
695	Yarn of acrylic staple fibres and cotton, not retail	17.8	6.0
696	Thread rolling machines for working metal, etc.	17.7	−2.7
697	Womens, girls blouses and shirts, material nes, not knit	17.6	−25.0
698	Womens, girls suits, of material nes, knit	17.6	−8.1
699	Electric capacitors, variable or adjustable (pre-set)	17.6	−1.5
700	Hats, plaited or made by assembling strips of material	17.5	−7.7
701	Paper, tarred, bituminised or asphalted, nes	17.4	−4.5
702	Mens, boys jackets and blazers, of cotton, not knit	17.4	−64.9
703	Hydrocarbon derivs, mixed sulpho/nitro/nitroso groups	17.3	−25.5
704	Used or new rags textile material, not sorted	17.3	1.6
705	Saddlery and harness, of any material	17.2	11.8
706	Pedestrian controlled tractors	17.1	0.5
707	Woven cotton nes, >85 % >200 g/m^2, yarn dyed, nes	17.1	5.5
708	Woven cotton nes, >85 % >200 g/m^2, printed, nes	17.1	10.0
709	Denim cotton nes, <85 % + manmade fibre, >200 g/m^2	17.1	−5.6

(continued)

Table 3.12 (continued)

Rank	Commodity description	Italy trade balance (million $)	Germany trade balance (million $)
710	Hot rolled stainless steel sheet, w <600 mm, t >4.75 mm	16.9	−35.0
711	Woven cotton fabric, >200 g/m^2, dyed, nes	16.9	−0.1
712	Plywood, sheets of tropical wood	16.8	−55.9
713	Multiple loudspeakers, mounted in single enclosure	16.8	−114.0
714	Mens, boys jackets and blazers, wool or hair, knit	16.8	−3.6
715	Semi-finished prod steel	16.6	−69.4
716	Sets of hand tools	16.6	13.9
717	Woven fabric polyester + wool or hair, nes	16.5	11.7
718	Cellulose acetates, plasticised, in primary forms	16.5	1.1
719	Air gun pellets, parts of shotgun cartridges	16.5	3.8
720	Woven fabric, combed wool or hair + manmade filament	16.4	−1.3
721	Ketone-alcohols nes; ket	16.4	−1.7
722	Snow-ski equipment except skis/bindings	16.4	−7.1
723	Woven fabric synthetic staple fibres, nes	16.3	4.0
724	Hand-operated appliances, food preparation, <10 kg	16.2	−11.5
725	Fertilizers with nitrates and phosphates, nes, <=10 kg	16.2	−28.3
726	Watch straps etc., and parts, of leather/plastic/etc.	16.2	−15.8
727	Mens, boys anoraks etc., of material nes, not knit	16.1	−1.1
728	Womens, girls jackets and blazers, material nes, knit	16.1	−9.9
729	Inner tubes of rubber for motor vehicles	16.0	−4.1
730	Corsets, braces and parts thereof	16.0	−17.5
731	Womens, girls skirts, synthetic fibres, not knit	15.9	−29.6
732	Wooden frames for paintings, photographs, mirrors etc.	15.8	−52.6
733	Woven cotton nes, >85 % <200 g/m^2, dyed	15.8	13.0
734	Blankets (non-electric) and travelling rug, wool or hair	15.8	3.2
735	Parts of presses, crushers for wine, fruit juice etc.	15.7	2.4
736	Carboxylic acids (aldehyde, ketone function), derivs	15.7	−6.2
737	Artist, student and signwriter colours, small packs nes	15.6	8.3
738	Mens, boys overcoats of material nes, not knit	15.6	−5.5
739	Mens, boys trousers and shorts, of cotton, not knit	15.5	−1,255.8
740	Mens, boys overcoats of manmade fibres, not knit	15.5	−56.4
741	Woven cotton nes, <85 % + manmade fibre, <200 g/m^2 dyed	15.4	0.6
742	Hand pumps not designed to measure flow	15.4	11.5
743	Aluminium pipe or tube fittings	15.4	4.0
744	Woven fabric >85 % nylon, polyamide, printed, nes	15.3	8.8

(continued)

Table 3.12 (continued)

Rank	Commodity description	Italy trade balance (million $)	Germany trade balance (million $)
745	Gold/silversmith wares of/clad with precious metal nes	15.3	4.3
746	Woven fabric >85 % artif filament/strip, yarn dyed, nes	15.2	1.9
747	Lead waste or scrap	15.1	−27.2
748	Lathes nes for removing metal	15.0	−4.8
749	Womens, girls anoraks etc., of wool or hair, not knit	14.9	−11.0
750	Woven fabric of artificial filament, yarn dyed, nes	14.9	1.6
751	Clothing accessories nes, of textile material, knit	14.8	−16.4
752	Babies garments, accessories of material nes, not knit	14.8	−2.1
753	Penicillins, derivatives, in bulk, salts	14.6	−16.9
754	Woven fabric synthetic filament <85 % + cotton, dyed nes	14.6	−1.3
755	Fittings, pipe and tube, nickel	14.6	7.7
756	Twill weave cotton, >85 % <200 g/m^2, yarn dyed	14.4	−5.0
757	Twill weave cotton, >85 % >200 g/m^2, bleached	14.2	3.7
758	Woven warp pile cotton, cut, except terry, w > 30 cm	14.2	2.0
759	Woven fabric <85 % artif staple + manmade fil yarn dyed	14.2	−2.8
760	Woven cotton nes, <85 % + manmade fibre, >200 g/m^2, dyed	14.1	−3.6
761	Woven fabric synthetic filament <85 % + cotton, yarn dye	14.1	−2.1
762	Engines, spark-ignition reciprocating, <50 cc	14.0	0.0
763	Bulldozers and angledozers, wheeled	13.8	−4.4
764	Other flat, in coils, not further worked than hot-rolled, pickled >4.75	13.8	−143.2
765	Woven fabric >85 % synthetic filament, yarn dyed, nes	13.8	−0.2
766	Woven fabrics <30 cm pla	13.8	−24.8
767	Pocket-watch, precious-metal case, non-battery	13.4	1.0
768	Antibiotics nes, formulated, in bulk	13.3	6.0
769	Woven fabric, carded wool or hair with natural fibre	13.2	1.8
770	Textile fibre carding machines	13.1	−1.8
771	AC generators, of an output 375–750 kVA	13.1	−12.9
772	Yarn of carded fine animal hair, not retail	13.1	−1.9
773	Plain weave cotton, >85 % >200 g/m^2, dyed	13.1	0.5
774	Woven fabric <85 % artificial staple + cotton, yarn dyed	13.1	1.7
775	Super-heated water boilers	13.0	11.6
776	Woven cotton fabric, >200 g/m^2, printed, nes	12.9	2.9
777	Mens, boys ensembles, of material nes, knit	12.8	−0.1

(continued)

Table 3.12 (continued)

Rank	Commodity description	Italy trade balance (million $)	Germany trade balance (million $)
778	Womens, girls jackets, blazers, synthetic fibres, knit	12.7	−33.2
779	Calcined dolomite	12.7	−27.7
780	Motorcycles, spark ignition engine of 250–500 cc	12.7	−31.3
781	Conveyor belts nes	12.7	−6.7
782	Headgear nes, of other materials	12.6	−7.4
783	Womens, girls blouses and shirts, of material nes, knit	12.5	−5.6
784	Womens, girls ensembles, of material nes, knit	12.3	−0.3
785	Mens, boys trousers and shorts, of material nes, knit	12.3	−4.5
786	Bentonite	12.3	−59.2
787	Yarn >85 % polyester staple fibres, multiple, not retail	12.3	−14.3
788	Worked human hair, wool or animal hair, for wig making	12.2	−4.4
789	Electric sound amplifier sets	12.2	10.7
790	Woven cotton fabric, <200 g/m^2, yarns mixed colours	12.2	−0.3
791	Glazed ceramic mosaic tiles, cubes and similar <7 cm wide	12.0	−0.2
792	Other flat, not in coils, not further worked than hot-rolled, with patterns in relief	12.0	−31.7
793	Condensers for steam or vapour power units	12.0	2.9
794	Mens, boys shirts, of material nes, not knit	11.9	−8.2
795	Scouring pastes and powders and other scouring preps	11.9	−2.9
796	Tools, broom handles, bodies, etc., of wood	11.8	−18.5
797	Rubber waste, parings and scrap (except hard rubber)	11.8	−4.1
798	Label, badge, etc., non-embroidered textile, not woven	11.8	4.0
799	Cotton yarn (except sewing thread) <85 % cotton, retail	11.7	−1.5
800	Sulphites of metals other than sodium	11.7	−0.3
801	Flat, not in coils, not further worked than cold-rolled (cold-reduced) <3 m	11.7	−1.8
802	Woven nes cotton, <85 % + manmade fibre, <200 g yarn dyed	11.7	−0.3
803	Yarn >85 % nylon, etc., staple fibres, single, not retail	11.6	−7.7
804	Vitamins A, derivatives, unmixed	11.5	−58.6
805	Molybdenum, articles thereof nes	11.5	−6.7
806	Other Ski-boots, cross-country ski footwear and snowboard boots	11.5	0.1
807	Woven fabric acrylic or modacrylic staple fibres, nes	11.4	−1.8
808	Yarn <85 % synthetic staple fibres, retail not sewing	11.3	−7.3
809	Woven fabric of noil silk	11.3	−0.4
810	Waste and scrap of prim cell	11.2	−8.5

(continued)

Table 3.12 (continued)

Rank	Commodity description	Italy trade balance (million $)	Germany trade balance (million $)
811	Conveyor belt textil m r	11.2	−91.0
812	Ice skates, roller skates, skating boots	11.2	−34.1
813	Textile fabric treated for theatrical backdrops etc.	11.1	−10.2
814	Hats and other headgear, knit or crochet, nes	11.1	−128.0
815	Yarn of viscose rayon, multiple, nes, not retail	11.1	−2.9
816	Smoking pipes and pipe bowls	11.0	3.3
817	Mens, boys trousers and shorts, material nes, not knit	11.0	−24.1
818	Amino-aldehydes, ketones and quinones, salts thereof	10.9	3.9
819	Inflatable balls	10.9	−25.3
820	Twill weave cotton, <85 % + manmade fibre, <200 g/m^2 dyed	10.9	0.0
821	Woven fabric >85 % artif filament/strip, printed, nes	10.7	1.7
822	Gun, musical instrument, camera, etc., cases, nes	10.7	−21.0
823	Tyre cord fabric of nylon, polyamides	10.7	−46.7
824	Sheep or lamb skins, raw, except pickled, no wool	10.6	0.0
825	Table/kitchen articles, parts, enamelled iron or steel	10.6	9.5
826	Mowers, powered, lawn, nes	10.5	−15.8
827	Textile wall coverings	10.5	8.3
828	Net/fencing, iron or steel <3 mm wire, <100 cm mesh, nes	10.4	5.8
829	Angles, shapes and sections, of iron or non-alloy steel, cold formed or cold finished and further worked	10.4	−9.1
830	Vegetable tanning extracts nes, tannins, salts, derivs	10.4	−3.5
831	Screw hooks and screw rings of iron or steel	10.4	4.8
832	Potassium hydroxide (caustic potash)	10.4	−49.7
833	Other Automatic beverage-vending machines	10.3	0.0
834	Woven fabric acrylics + manmade filament, nes	10.2	2.8
835	Quilted textile products in piece (not embroidered)	10.2	7.9
836	Animal carving material, articles, nes	10.2	−2.1
837	Ornamental trimmings in the piece	10.1	1.8
838	Womens, girls ensembles, synthetic fibres, not knit	10.1	−2.2
839	Non-conifer wood continuously shaped along any edges	9.9	−107.9
840	Woven cotton fabric, >200 g/m^2, printed, nes	9.9	−0.2
841	Calcium chloride	9.9	−14.6
842	Domestic iron/steel liquid/solid fuel appliances nes	9.9	−18.5
843	Waterproof footwear (Wellingtons etc.,), metal toe cap	9.8	−5.6
844	Table linen, of cotton, not knit	9.8	−44.8

(continued)

Table 3.12 (continued)

Rank	Commodity description	Italy trade balance (million \$)	Germany trade balance (million \$)
845	Evaporative air coolers	9.8	1.0
846	Flax yarn multiple (folded) or cabled	9.7	0.0
847	Powders, aluminium, of non-lamellar structure	9.7	−31.7
848	Woven fabric <85 % artificial staple fibres, dyed	9.6	0.6
849	Trucks nes	9.6	−4.9
850	Womens, girls anoraks etc., of material nes, not knit	9.5	−4.8
851	Zinc oxide and peroxide	9.4	0.5
852	Cream separators	9.3	0.9
853	Parts and accessories for the musical instruments nes	9.2	−12.5
854	Calcium carbonate	9.2	−44.6
855	Woven fabric polyester staple fibres, nes	9.1	1.1
856	Cutlery not in sets, not plated with precious metal	9.1	−25.2
857	Woven fabric, carded wool or hair + manmade filament	9.0	−0.2
858	Rivets, for clothing, footwear, bags/etc.	9.0	−4.6
859	Cotton yarn <85 % multiple uncombed >714, not ret., nes	9.0	−8.1
860	Womens, girls trousers and shorts, of wool hair, knit	8.9	−0.2
861	Extracts of glands etc., for therapeutic use	8.9	2.6
862	Muzzle-loading firearms	8.8	−0.4
863	Woven fabric of hi-ten filament yarns of viscose rayon	8.8	−0.1
864	Watch cases of base metal including gold/silver-plated	8.8	−14.4
865	Twill cotton except denim, >85 % >200 g/m^2, yarn dyed	8.8	2.0
866	Woven fabric synthetic staple fibre with manmade, nes	8.7	0.6
867	Aircraft engines, spark-ignition	8.6	−3.0
868	Wire bars, copper, unwrought	8.6	−0.3
869	Woven fabric materials nes, <30 cm wide	8.6	2.3
870	Cotton, carded or combed	8.4	−2.0
871	Drilling, threading or tapping tools	8.4	4.7
872	Unit construction machines, metal work	8.3	7.7
873	Woven fabric <85 % artificial staple + wool or hair, dyed	8.3	0.7
874	Woven fabric >85 % synth nes + cotton, >170 g/m^2 yarn dye	8.3	5.1
875	Plain weave cotton, <85 % + manmade fibre, <200 g/m^2 dyed	8.2	4.7
876	Twill weave cotton, >85 % <200 g/m^2, bleached	8.1	3.1
877	Articles for pocket or handbag, nes	8.0	−6.7

(continued)

Table 3.12 (continued)

Rank	Commodity description	Italy trade balance (million $)	Germany trade balance (million $)
878	Penicillins or streptomycins and derivatives, in bulk	8.0	−0.8
879	Globes, geographical, printed	8.0	−0.8
880	Flat, plated or coated with tin, w >600 mm, t >0.5 mm	8.0	−0.6
881	Bicycle wheel rims and spokes	8.0	−20.5
882	Tents, of textile material nes	7.9	0.8
883	Yarn, polyester, single, >50 turn/m, not retail	7.9	−17.5
884	Polyhydric acyclic alcohols nes	7.8	−4.9
885	Waste, scale, dross, slag of iron or steel industry	7.8	−70.6
886	Woven fabric >85 % nylon, polyamide, yarn dyed, nes	7.7	−0.6
887	Track suits, of materials nes, knit	7.7	−1.2
888	Carpets and textile floor coverings, nes	7.7	−24.9
889	Sharpening (tool, cutter grind) machines nes	7.7	5.4
890	Particle board of ligneous materials except wood	7.6	0.9
891	App and equip for ph labor	7.6	0.8
892	Lamp and lighting fitting parts of plastics	7.6	6.8
893	Pile knit or crochet fabric, of manmade fibres, nes	7.5	1.2
894	Hexamethylenediamine, its salts	7.4	−636.5
895	Office supplies nes, of paper, book covers, blotters	7.2	4.0
896	Horizontal lathes nes for metal	7.1	4.2
897	Walking-sticks, seat-sticks, whips, etc.	7.1	−8.0
898	Parts of lighters, other than flints/wicks	7.1	−1.5
899	Woven cotton nes, <85 % + manmade fibre, <200 g/m^2 print	7.0	0.0
900	Projection screens	7.0	−11.9
901	Methanal (formaldehyde)	6.9	−44.3
902	Woven fabric <85 % artif staple + wool or hair, yarn dyed	6.9	−0.2
903	Yarn of acrylic staple fibres, not retail, nes	6.8	0.8
904	Wirewound variable resistors, rheostats, etc., >20 W	6.8	3.1
905	Woven fabric of other vegetable textile fibre, paper	6.8	−1.9
906	Woven fabric <85 % artificial staple fibre, yarn dyed	6.8	−0.3
907	Monumental or building stone nes, porphyry and basalt	6.8	−2.6
908	Womens, girls swimwear, not knit	6.8	−2.5
909	Sheet/film not cellular/reinf acrylic polymers nes	6.8	−24.1
910	Yarn of other synthetic staple with wool or hair, nes	6.8	−0.2
911	Albumins nes, albuminates and other albumin derivatives	6.7	1.6
912	Handkerchiefs, of material nes, not knit	6.7	−0.9

(continued)

Table 3.12 (continued)

Rank	Commodity description	Italy trade balance (million $)	Germany trade balance (million $)
913	Paper, sensitising base stock	6.7	−4.8
914	Thiuram mono-, di- or tetra-sulphides	6.7	−4.6
915	Tarpaulins, awnings and sunblinds, of material nes	6.6	−9.8
916	Pile knit, crochet fabric, natural yarn not cotton nes	6.5	−0.7
917	Ultrasonic scanning appr	6.5	−49.4
918	Loop pile knit/crochet fabric, natural yarn not cotton	6.4	−0.2
919	Tube/pipe nes, iron/steel riveted etc., diameter >406.4 mm	6.4	0.0
920	Bed linen, of material nes, printed, not knit	6.4	−1.8
921	Woven fabric <85 % artificial staple + cotton, dyed	6.4	−0.8
922	Textured yarn, synthetic filament, nes, not retail	6.3	−8.8
923	Twill weave cotton, >85 % >200 g/m^2, printed	6.3	4.6
924	Angles, shapes and sections, welded, iron or steel	6.3	−7.8
925	Womens, girls suits, synthetic fibres, knit	6.2	−2.0
926	Clocks, nes, non-electric	6.1	4.5
927	Colouring matter of vegetable or animal origin	6.1	−12.7
928	Bricks, blocks and ceramic goods of siliceous earths	6.1	−3.6
929	Cuff-links and studs of base metal, plated or not	6.1	5.9
930	Twill weave cotton, >85 % >200 g/m^2, unbleached	6.1	−13.1
931	Ampoules of glass for conveyance or packing	6.0	3.6
932	Woven fabric <85 % artificial staple + wool/hair, printed	6.0	−0.1
933	Granite, merely cut into blocks etc.	5.9	−12.1
934	Womens, girls skirts, of material nes, knit	5.8	−5.4
935	Mens, boys trousers, shorts, of synthetic fibres, knit	5.8	−30.0
936	Wire of silico-manganese steel	5.8	−19.3
937	Parts of liquid elevators	5.8	3.0
938	Hammers including sledge hammers	5.7	−5.1
939	Portland cement, white or white artificially coloured	5.7	−9.8
940	Bases, moulding	5.7	−12.0
941	Yarn >85 % acrylic staple fibres, multiple not retail	5.7	−5.3
942	Felt, not of wool or hair, not impregnated or coated	5.6	2.7
943	Iron or non-alloy steel in ingots, <99.94 % iron	5.6	−21.1
944	Boxes, pouches, wallets, etc., of stationary	5.6	−1.4
945	Rhodium unwrought or in powder form	5.6	−52.4
946	Cutlery sets plated with precious metal	5.5	3.8
947	Womens, girls blouses and shirts, wool or hair, not knit	5.5	−2.1

(continued)

Table 3.12 (continued)

Rank	Commodity description	Italy trade balance (million $)	Germany trade balance (million $)
948	Shuttle type looms nes for weaving fabric >30 cm wide	5.5	0.1
949	Refractory ceramic articles, >50 % carbon	5.5	−0.8
950	Statuettes and other ornaments of wood	5.5	−53.2
951	Bed linen, of material nes, nes	5.4	0.8
952	Rubber thread and cord, textile covered	5.4	1.0
953	Set, woven fabric and yarn for rugs, tapestry etc.	5.4	0.7
954	Cotton yarn >85 % single uncombed >714 dtex, not retail	5.3	−12.0
955	Borates of metals except refined borax	5.3	−2.4
956	Woven cotton nes, >85 % >200 g/m^2, bleached, nes	5.3	3.1
957	Conveyor belt metal rein	5.3	−27.6
958	Salts and esters of tartaric acid	5.2	−2.2
959	Made up fishing nets, of manmade textile materials	5.2	−1.3
960	Twill cotton not denim <85 % + manmade fibre, >200 g y-dyed	5.2	−0.2
961	Woven cotton nes, <85 % + manmade fibre, >200 g, bleached	5.2	−2.7
962	Gum or amylaceous covered textiles for book covers etc.	5.1	2.3
963	Yarn synthetic filament, multiple, nes, not retail	5.1	0.0
964	Floating docks, special function vessels nes	5.1	−408.1
965	Safety or detonating fuses, detonators, igniters	5.1	−33.1
966	Sacks and bags, packing, of cotton	5.0	−3.2
967	Organic compounds, nes	5.0	−10.7
968	Tracing cloth, painting canvas, stiffened textile nes	5.0	−5.3
969	Tulles, other nets (not woven, knit or crochet)	4.9	−7.3
970	Railway fish plates and sole plates, iron or steel	4.9	−2.6
971	Cloth, grill, aluminium wi	4.9	3.5
972	Panels, laminated wood, nes	4.8	−15.7
973	Asbestos fabricated products nes	4.8	0.0
974	Reaction engines other than turbo jets	4.7	−602.0
975	Alkaloids, derivs, without antibiotics, hormones, bulk	4.6	−2.7
976	Casings,circular, iron/steel, oil/gas drilling >406 mm	4.6	2.3
977	Oxygen	4.6	−15.1
978	Net/fencing, galvanized i/nas <3 mm wire, <100 cm mesh	4.6	−2.6
979	Parts of electronic valve and tubes, except cathode ray	4.6	−4.2
980	Coin (other than gold coin) not being legal tender	4.5	−368.6
981	False beard, eyebrows etc., synthetic textile material	4.5	−5.5
982	Waterproof footwear(Wellington) no toe cap, nes	4.5	−2.6

(continued)

Table 3.12 (continued)

Rank	Commodity description	Italy trade balance (million $)	Germany trade balance (million $)
983	Hypochlorites nes, chlorites and hypobromites of metals	4.5	0.6
984	Acetone	4.5	−61.6
985	Cotton yarn <85 % single uncombed >714 dtex, not retail	4.5	−3.0
986	Parts for hats and headgear	4.4	−4.1
987	Household and toilet articles nes of porcelain or china	4.4	−0.6
988	Nickel waste or scrap	4.3	−51.6
989	Woven pile, chenille fabric of yarn nes, except terry	4.3	−0.4
990	Woven fabric <85 % artificial staples, printed, nes	4.3	−1.2
991	Boring-milling machines nes for removing metal	4.3	−0.4
992	Cargo containers designed for carriage	4.2	−46.5
993	Woven fabric >85 % synth nes + cotton, <170 g/m^2 yarn dye	4.2	−0.4
994	Unglazed ceramic mosaic tiles etc., <7 cm wide	4.2	0.8
995	Unroasted iron pyrites	4.1	−6.5
996	Womens, girls bathrobe, dressing gowns, nes, knit	4.1	1.5
997	Womens, girls jackets, blazers, synth fibres, not knit	4.0	−117.4
998	Yarn of artificial fibres except sewing thread, retail	4.0	−0.6
999	Yarn >85 % other synth staple fibres, single not retail	4.0	2.7
1000	Toilet or kitchen linen, of material nes	3.9	−6.2
1001	Bulldozer and angledozer blades	3.9	3.5
1002	Mechanical lace, other material (piece, strip, motif)	3.9	3.5
1003	Polishes, creams etc., for maintenance of woodwork	3.9	2.8
1004	Torches, hand-held, for soldering/brazing/welding	3.9	−1.0
1005	Wrist-watch, precious metal, battery, with hands	3.9	−48.7
1006	Articles of porcelain or china nes	3.8	−0.3
1007	Womens, girls garments nes, of cotton, not knit	3.8	−55.9
1008	Artificial teeth	3.7	−7.1
1009	Pneumatic tyres used	3.6	0.9
1010	Yarn of other vegetable textile fibres	3.6	0.7
1011	Woven fabric <85 % artificial staple + manmade fibre dyed	3.6	1.2
1012	Womens, girls skirts, of wool or hair, knit	3.6	−1.0
1013	Pneumatic hand tool parts	3.5	−7.8
1014	Yarn of other synth staple fibres + cotton not retail	3.5	3.2
1015	Paper, felt, uncoated	3.5	−5.7
1016	Essential oils of orange	3.5	−33.8
1017	Shaving, hair, nail, eyelash and other toilet brushes	3.5	−6.5
1018	Glassware except kitchen, table ware, of lead crystal	3.4	3.2

(continued)

Table 3.12 (continued)

Rank	Commodity description	Italy trade balance (million $)	Germany trade balance (million $)
1019	Womens, girls swimwear, of material nes, knit	3.4	−1.2
1020	Hot rolled stainless steel sheet, w < 600 mm, t < 4.75 mm	3.4	−22.0
1021	Parts and accessories nes of weapons, nes	3.4	−9.8
1022	Ignition magnetos, magneto-generators and flywheels	3.3	−14.0
1023	Chain, stud link, iron or steel	3.3	1.5
1024	Wooden cases, boxes, crates, drums and containers	3.2	−46.7
1025	Mens, boys garments nes, of manmade fibres, not knit	3.2	−48.4
1026	Plain weave cotton, <85 % + manmade fibre, >200 g, yarn dyed	3.2	0.0
1027	Silver nitrate	3.2	−38.0
1028	Wigs, false beards, eyebrows etc., of human hair	3.1	−6.1
1029	Woven fabric of flax, <85 % flax, unbleached/bleached	3.1	−0.2
1030	Greasy wool (other than shorn) not carded or combed	3.0	1.5
1031	Plain weave cotton, <85 % + manmade fibre, <200 g print	3.0	−0.8
1032	Knotted netting, nets, of of natural materials	2.9	−0.4
1033	Womens, girls overcoats, etc., material nes, knit	2.9	−6.7
1034	Warp knit fabric of other materials, nes	2.9	−1.0
1035	Letterpress printing machinery nes except flexographic	2.9	−8.0
1036	Machines for weaving fabric, width <30 cm	2.9	−0.1
1037	Cotton sewing thread >85 % cotton, not retail	2.9	0.0
1038	Household/laundry-type washing machines <10 kg, nes	2.8	−1.8
1039	Harpsichords, keyboard stringed instruments nes	2.8	0.2
1040	Nitrogen	2.8	−9.4
1041	Fishing vessels and factory ships	2.7	0.3
1042	Gloves, mittens or mitts, nes, of wool or hair, knit	2.7	−4.9
1043	Mens, boys suits, of materials nes, knit	2.7	−1.9
1044	Plain weave cotton, >85 % >200 g/m^2, yarn dyed	2.7	−1.5
1045	Womens, girls suits, synthetic fibres, not knit	2.6	−0.4
1046	Yarn >85 % synthetic staple fibres, retail, not sewing	2.6	−21.2
1047	Leather, composition sports gloves, mittens and mitts	2.6	−4.7
1048	Cotton yarn <85 % multiple uncomb 714–232 dtex, not ret.	2.6	−0.2
1049	Maps and charts, printed, in book form	2.6	−1.0
1050	Typewriters, electric, nes	2.6	−0.4
1051	Umbrella frames	2.6	−2.3

(continued)

Table 3.12 (continued)

Rank	Commodity description	Italy trade balance (million $)	Germany trade balance (million $)
1052	Cotton yarn <85 % multiple combed >714, not retail, nes	2.6	−0.6
1053	Textile fibre combing machines	2.5	0.0
1054	Woven cotton nes, <85 % + manmade fibre, <200 g bleached	2.4	2.1
1055	Woven fabric incorporating metal threads, nes	2.4	−0.5
1056	Ephedrines and their salts	2.4	−0.3
1057	Looped pile knit or crochet fabric, of cotton	2.4	−3.7
1058	Plain weave cotton, <85 % + manmade fibre, >200 g, dyed	2.4	−0.9
1059	Carbon tetrachloride	2.4	1.6
1060	Candles, tapers and the like	2.3	−222.7
1061	Cotton yarn >85 % single uncombed <125 dtex, not retail	2.3	−0.1
1062	Sails, synthetic fibres	2.3	−6.9
1063	Woven fabric synthetic filament, <85 % + cotton, printed	2.2	0.9
1064	Woven fabric >85 % synthetic nes + cotton, >170 g/m^2 dyed	2.2	0.8
1065	Original engravings, prints and lithographs	2.1	−3.0
1066	Sulphur, sublimed or precipitated, colloidal sulphur	2.1	−9.1
1067	Prepared driers	2.1	−9.1
1068	Articles of precious, semi-precious, artificial stones	2.1	−19.6
1069	Curtains drapes blinds valances, material nes, woven	2.1	−3.9
1070	Fishing and hunting equipment and requisites nes	2.1	−9.6
1071	Ash or residues containing mainly aluminium	2.1	−97.6
1072	Ski suits, of textile material, not knit	2.1	−11.0
1073	Yarn >85 % other synth staple fibre multiple not retail	2.0	−0.9
1074	Plain weave cotton, >85 % >200 g/m^2, printed	2.0	−1.6
1075	Mens, boys trousers and shorts, of wool or hair, knit	2.0	−4.0
1076	Table, kitchen or other	2.0	0.9
1077	Mens, boys jackets and blazers, synthetic fibres, knit	2.0	−13.7
1078	Lamp and lighting fitting parts of glass	2.0	−2.8
1079	Curtains drapes blinds valances, material nes, knit	2.0	−0.3
1080	Cotton yarn <85 % multiple combed <125 dtex, not retail	1.9	−1.1
1081	Womens, girls slips etc., of material nes, not knit	1.9	−1.1
1082	Work truck parts	1.8	−0.3
1083	Photo plates and film, exposed and developed, for offset	1.8	−0.5
1084	Cutlery sets, articles, not plated with precious metal	1.8	−15.9
1085	Stones for milling, grinding or pulping	1.8	−1.1
1086	Scythes, sickles etc., used in agriculture, etc.	1.8	−9.8

(continued)

Table 3.12 (continued)

Rank	Commodity description	Italy trade balance (million $)	Germany trade balance (million $)
1087	Axes, bill hooks and similar hewing tools	1.8	−6.8
1088	Mens, boys ensembles, synthetic fibres, knit	1.8	−2.2
1089	Yarn of synthetic filament not sewing thread, retail	1.8	−1.7
1090	Glass-ceramic ware for table kitchen, etc.	1.8	−2.0
1091	Ash or residues containing mainly lead	1.7	−170.3
1092	Babies garments, accessories of material nes, knit	1.7	0.0
1093	Polyvinyl chloride waste or scrap	1.7	−13.5
1094	Milling machines, knee-type nes for removing metal	1.7	−0.2
1095	Chain parts, articulated link, iron or steel	1.7	−1.6
1096	Plain weave cotton <85 % + manmade fibre, <200 g bleached	1.7	−0.9
1097	Cotton yarn <85 % multiple combed 714–232 dtex, not ret.	1.6	0.1
1098	Yarn, cellulose acetate, multiple, nes, not retail	1.6	0.0
1099	Reclaimed rubber in primary forms or in sheets	1.6	−1.1
1100	Woven fabric <85 % artif staple + manmade fil, printed	1.6	0.2
1101	Signalling glassware, unworked optical elements	1.6	−29.6
1102	Cotton yarn >85 % multiple combed >714 dtex, not retail	1.6	0.2
1103	Parts and accessories of electronic calculators	1.5	−2.6
1104	Electric capacitors, fixed, nes	1.5	−18.0
1105	Sacks and bags, packing, of materials nes	1.5	−2.2
1106	Cotton yarn >85 % multiple uncombed >714 dtex, not ret.	1.5	−1.6
1107	Woven cotton nes, <85 % + manmade fibre, >200 g/m^2 unbl	1.5	1.0
1108	Millstones, grindstones etc., of natural stone	1.5	−0.4
1109	Yarn synth filament, single, >50 turn/m nes not retail	1.4	−8.4
1110	Mens, boys overcoats, etc., of manmade fibres, knit	1.4	−35.3
1111	Wrench sockets, ratchet handles, extensions, spinners	1.4	−8.5
1112	Mens, boys underpants or briefs, of cotton, not knit	1.4	−30.1
1113	Sailboards	1.4	−4.2
1114	Cotton yarn (except sewing thread) >85 % cotton, retail	1.4	−7.5
1115	Woven plain >85 % polyester + cotton, >170 g/m^2 dyed	1.4	−0.1
1116	Dolomite not calcined	1.3	−1.2
1117	Womens, girls ensembles, synthetic fibres, knit	1.3	−5.8
1118	Cameras for preparing printing plates or cylinders	1.3	0.2
1119	Hand printing, dating and numbering devices, etc.	1.3	−8.2

(continued)

Table 3.12 (continued)

Rank	Commodity description	Italy trade balance (million $)	Germany trade balance (million $)
1120	Woven cotton fabric, >200 g/m^2, bleached, nes	1.2	0.8
1121	Photographic plate, film, paper, exposed, undeveloped	1.2	−0.2
1122	Womens, girls slips or petticoats, material nes knit	1.2	−0.4
1123	Plain weave cotton, <85 % + manmade fibre, <200 g yarn dyed	1.2	0.4
1124	Cotton yarn <85 % single uncombed 232–192 dtex, not ret.	1.1	−0.2
1125	Brooms/brushes of vegetable material	1.1	−1.7
1126	Spades and shovels	1.1	−11.9
1127	Gloves, mittens or mitts, nes, material nes knit	1.1	−1.6
1128	Ties, bow ties and cravats, manmade fibres, not knit	1.1	−3.0
1129	Cotton yarn <85 % single uncombed 192–125 dtex, not ret.	1.1	0.1
1130	AC motors, single-phase, nes	1.1	−65.5
1131	Cotton sewing thread, <85 % cotton, not retail	1.0	−0.1
1132	Tools for working in the hand, pneumatic rotary type	1.0	−17.2
1133	Blankets (except electric) and travel rugs, material nes	1.0	−3.4
1134	Alloy pig iron, spiegele	1.0	0.0
1135	Woven fabric synthetic filament nes	1.0	0.6
1136	Watch movements, complete and assembled, nes	0.9	−2.0
1137	Cotton yarn <85 % single uncombed <125 dtex, not retail	0.9	0.0
1138	Hand sharpening or polishing stones	0.9	0.1
1139	Copper mattes	0.9	−13.9
1140	Vessels and other floating structures for breaking up	0.9	0.0
1141	Carded wool	0.9	0.3
1142	Woven cotton fabric, >200 g/m^2, unbleached, nes	0.9	0.0
1143	Baths, cast iron	0.9	0.2
1144	Cotton yarn >85 % multiple combed 714–232 dtex, not ret.	0.9	−0.4
1145	Clocks with watch movements, battery (except vehicle)	0.9	−1.9
1146	Womens, girls jackets and blazers, of cotton, knit	0.8	−42.5
1147	Yarn, viscose rayon, single >120 turn/m nes, not retail	0.8	−0.4
1148	Woven fabric of coarse animal hair or horse hair	0.8	−0.1
1149	Staple fibres of acrylic, modacrylic, carded or combed	0.8	−0.1
1150	Earth colours with >70 % iron as Ferrous oxide	0.8	−1.3
1151	Monochrome cathode-ray picture tubes, monitors	0.8	0.1
1152	Mens, boys underpants or briefs, material nes, knit	0.8	−0.6
1153	Girdles, panty girdles and parts thereof	0.8	−4.3
1154	Articles nes, of asbestoso or cellulose fibre cement	0.8	0.5
1155	Calculating machines, non-electric	0.7	0.0

(continued)

Table 3.12 (continued)

Rank	Commodity description	Italy trade balance (million $)	Germany trade balance (million $)
1156	Railway tank cars	0.7	−63.5
1157	Woven fabric of jute/bast fibre, not unbleach/bleached	0.7	−0.2
1158	Wigs, false beards, eyebrows etc., of other materials	0.7	−1.9
1159	Table/kitchen articles, parts, unenamelled cast iron	0.7	−0.8
1160	Molybdenum ores and concentrates except roasted	0.7	−0.9
1161	Furnishing goods nes, material nes, not knit, crochet	0.7	−8.3
1162	Woven fabric >85 % synth staple fibre nes unbl/bleached	0.7	−12.3
1163	Piezo-electric quartz	0.7	−2.1
1164	Parts and trimmings, of umbrella, walking sticks, etc.	0.7	−6.4
1165	Ergotamine (INN) and it	0.6	0.3
1166	Artif monofilament >67 dtex t <1 mm, strip, straws t <5 mm	0.6	−0.3
1167	Unexposed photographic paper, board or textile nes	0.6	−0.4
1168	Yarn of polyester staple fibres, not retail, nes	0.6	−0.5
1169	Woven fabric >85 % synthetic filament, printed, nes	0.6	0.0
1170	Roasted iron pyrites	0.6	−3.7
1171	Musical instruments etc., nes	0.6	0.3
1172	Antimony ores and concentrates	0.5	0.0
1173	Gauze, except cotton, >30 cm wide	0.5	0.0
1174	Cotton yarn <85 % multiple combed 192–125 dtex, not ret.	0.5	−0.7
1175	Natural borates etc., nes, natural boric acid (<85 %)	0.5	−2.5
1176	Potassic fertilizers, mixes, nes, pack >10 kg	0.5	−1.7
1177	Woven plain >85 % polyester + cotton, >170 g/m^2 unbl/blchd	0.5	−1.1
1178	Woven fabric >85 % synthetic nes + cotton, <170 g/m^2 dyed	0.5	0.3
1179	Bicycle brakes, parts thereof	0.5	−86.7
1180	Woven twill >85 % polyester + cotton, >170 g/m^2 printed	0.5	0.1
1181	Clocks with watch movements, nes (except vehicle)	0.4	−0.2
1182	Hand sieves and hand riddles	0.4	−2.3
1183	Float glass etc., in sheets, wired	0.4	−1.1
1184	Balls nes	0.4	−13.9
1185	Anthraquinone	0.4	0.0
1186	Graphite based products nes	0.3	−3.5
1187	Twine nes, cordage, ropes and cables, of sisal	0.3	−1.0
1188	Clock, etc., cases, except metal	0.3	−0.5
1189	Copper ores and concentrates	0.3	−2,939.1
1190	Derivatives of phenols or phenol-alcohols, nes	0.3	−2.8
1191	Carpets of materials nes, knotted	0.3	−8.7

(continued)

Table 3.12 (continued)

Rank	Commodity description	Italy trade balance (million $)	Germany trade balance (million $)
1192	Rare-earth metals, scandium and yttrium	0.3	−16.2
1193	Naphthols, salts	0.3	−1.3
1194	Pocket-watch, precious-metal case, battery	0.3	0.0
1195	Clock movements, un/partly assembled, rough clocks	0.2	−2.6
1196	Letter corners, paper clips, metal office articles nes	0.2	−6.9
1197	Cobalt ores and concentrates	0.2	−0.8
1198	Plain weave cotton, <85 % + manmade fibre, >200 g, print	0.2	−0.4
1199	Lumber, White Lauan, Whi	0.2	−0.2
1200	Yarn, nylon, polyamide, single >50 turn/m, not retail	0.2	−0.2
1201	Womens, girls garments nes, manmade fibres, not knit	0.2	−123.8
1202	Curtains drapes blinds valances, synthetic fibre, knit	0.2	−32.3
1203	Pituitary anterior hormones and derivatives, in bulk	0.2	−275.1
1204	Woven uncut weft pile fabric of manmade fibre	0.1	−0.2
1205	Carpets of wool or fine hair, woven, made up, nes	0.1	−7.7
1206	Precious metal ores and concentrates except silver	0.1	−488.2
1207	Commercial alkali metal silicates, silicates nes	0.1	−0.4
1208	Mercuric oxide primary c	0.1	0.0
1209	Mica in crude form, sheets and splittings	0.1	−5.2
1210	Balata, gutta-percha, guayule, chicle and similar gums	0.1	0.0
1211	Bromochlorodifluorometha	0.1	0.0
1212	Cinematograph film, exposed and developed, width <=35 mm	0.1	0.0
1213	Mens, boys anoraks etc., of cotton, not knit	0.1	−77.7
1214	Natural uranium, its compounds, mixtures	0.1	−514.4
1215	Cotton yarn <85 % single combed >714 dtex, not retail	0.1	0.0
1216	Planes, chisels, gouges and similar wood cutting tools	0.1	−0.8
1217	Tantalum unwrought, bars, rods simply sintered, scrap	0.1	−45.0
1218	Flax fibre, broken or scutched	0.1	0.0
1219	Carpets of other manmade textile materials, tufted	0.1	−141.3
1220	Waste of coarse animal hair, except garnetted stock	0.1	0.0
1221	Prepared rubber accelerators	0.1	−15.5
1222	Woven fabric >85 % acrylic staple fibres, unbl/bleached	0.1	−0.5
1223	Cadmium, articles thereof, nes	0.0	−0.1
1224	Jute and other bast fibres, not spun, nes, tow, waste	0.0	−2.0
1225	True hemp fibre otherwise processed but not spun	0.0	−1.7
1226	Natural graphite, except powder or flakes	0.0	−1.5
1227	Woven plain >85 % polyester + cotton, >170 g/m^2 printed	0.0	−0.2

(continued)

Table 3.12 (continued)

Rank	Commodity description	Italy trade balance (million $)	Germany trade balance (million $)
1228	Aminohydroxynaphthalenesulphonic acids and salts	0.0	−1.0
1229	Beryllium, unwrought, waste or scrap/powders	0.0	0.0
1230	Combed wool in fragments	0.0	−2.1
1231	Silk-worm cocoons suitable for reeling	0.0	0.0
1232	Lead powders and flakes	0.0	−0.4
1233	Coarse animal hair, carded or combed	0.0	0.0
1234	Terpenic oils nes, crude dipentene, paracymene	0.0	−0.3
1235	Incomplete movements of watches, assembled	0.0	−0.3
Total		145,909.2	−17,027.0

Fortis-Corradini Index, © Fondazione Edison. Analysis on the basis of about 4,000 manufactured products excluding food and energy

Source Compiled by Fondazione Edison on data from Istat, Eurostat and UN Comtrade

References

EUROSTAT, Eurostat Database, http://ec.europa.eu/eurostat

Fortis M (2013) L'Italia batte la Germania in 1.200 prodotti. «Quaderno di Approfondimenti statistici», Fondazione Edison, no. 121, January

Fortis M, Corradini S (2010) I mille primati del Made in Italy. «Quaderno di Approfondimenti Statistici», Fondazione Edison, no. 47, January

Fortis M, Quadrio Curzio A (eds) (2006) Industria e distretti. Un paradigma di perdurante competitività italiana. Fondazione Edison Series. Il Mulino, Bologna.

ISTAT, Statistiche del commercio estero, www.coeweb.istat.it

UN Comtrade, International Trade Statistics Database, http://comtrade.un.org

MIX
Papier aus verantwortungsvollen Quellen
Paper from responsible sources
FSC® C105338

If you have any concerns about our products,
you can contact us on
ProductSafety@springernature.com

In case Publisher is established outside the EU,
the EU authorized representative is:
Springer Nature Customer Service Center GmbH
Europaplatz 3, 69115 Heidelberg, Germany

Printed by Libri Plureos GmbH
in Hamburg, Germany